I0055786

Electrical Weeding

A Sustainable Alternative to Herbicides

Sergio de Andrade Coutinho Filho

CABI is a trading name of CAB International

CABI
Nosworthy Way
Wallingford
Oxfordshire OX10 8DE
UK

Tel: +44 (0)1491 832111
E-mail: info@cabi.org
Website: www.cabi.org

CABI
200 Portland Street
Boston
MA 02114
USA

Tel: +1 (617)682-9015
E-mail: cabi-nao@cabi.org

A catalogue record for this book is available from the British Library, London, UK.

ISBN-13: 9781836992264 (hardback)
9781836992271 (ePDF)
9781836992288 (ePub)

DOI: 10.1079/9781836992288.0000

Commissioning Editor: Ward Cooper
Editorial Assistant: Theresa Regueira
Production Editor: Rosie Hayden

Typeset by Straive, Pondicherry, India
Printed in the USA

Contents

Acknowledgments

I would like to express my deepest gratitude to those who have been fundamental to the journey and success of Zasso.

First and foremost, I acknowledge Satoru Narita and Sylvio Coutinho, whose vision and determination as co-founders laid the foundation for the company's innovation. Their pioneering spirit continues to be a source of inspiration.

A special thank you to Benjamin Ergas, my co-CEO, and his family, whose leadership, strategic insights, and unwavering commitment have been instrumental in driving Zasso forward.

I am also grateful for the invaluable support of our institutional partners, who have contributed significantly to the advancement and application of our technologies.

Zasso's success would not be possible without the dedication and expertise of the entire team, whose relentless pursuit of innovation and excellence fuels our progress every day.

Additionally, I extend my appreciation to our R&D partners around the world, whose collaboration and scientific contributions continue to enhance our technological capabilities and expand the impact of electrical weeding.

We are just beginning to redefine the weeding paradigm, and this technology is only scratching the surface of its potential. To everyone who has played a role in this journey, directly or indirectly, I offer my sincerest thanks. Your support and belief in our mission are truly invaluable.

Further, I would acknowledge:

- Zasso's engineers who have helped and participated in the developments that are here described in different ways: Adriano Andrade, Alan Hespanhol, Alex Manoel, Bruno Valverde, Christopher Freimann, Denilson Marques, Diego Sousa, Elvis Diniz, Erik Castro, Felipe Leme, Felipe Nunes, Franz Hopfinger, Guilherme Rinzler, Isabela Meneghin, Jeancarlo and Constantino Schwager, Johatan Araujo, Maximilian Jansen, Maximilian Koch, Paulo Pereira, Stefan Lidak, Stephan Nolden, The Phong Nguyen and so many others.
- External development partners with whom we have co-developed many different projects throughout the years. Among them, I would like to mention: Artur Martini da Rosa, Bruna Evelin Gomes, Carlos Ricardo Barbosa, Carlos Solon, Eduardo Chini de Freitas, Estevan Linck Lara, Felipe Arnhold, Giovanni Gaiardo, Giusepe Discondi Dellagrave, Leonardo Rocha, Kevyn Santos, Luiz Vicente, Maicon de Souza Alves, Maik Basso, Mario Lucio Martins, Pedro Schnardorf, Rafael Feijó, Rafael Trintinaglia, Vitor Nardelli, Taciano Ares Rodolfo, and Willian Sciacca.
- External research partners who have developed or with whom we have co-developed many different projects throughout the years. Among them, I would like to mention: Alexandre Magno Brighenti, Danieli Simonetti, Deodoro Magno Brighenti, Luiza Baccin, Lynn Sosnokie, Marcelo Moretti, Maurilio Oliveira, Michel Martin, Muthu V. Bagavathiannan, Rafael Pedroso, Ryan Hamberg, and Thierry Besancon.

1

Background

Why Control Plants?

Weeds are one of the major causes of crop losses on farms around the world. They can reduce crop yields by competing with the crops for essential resources such as water, nutrients, and sunlight.

The amount of crop loss due to weeds can vary depending on a number of factors. For example, certain crops may be more susceptible to weed competition than others. In addition to the type of crop, the type of weed can also play a role in the amount of crop loss due to weeds. Some weeds are more aggressive and can outcompete crops more easily, leading to higher levels of crop loss. In some cases, certain weed species may also be more resistant to certain weed management techniques, making them more difficult to control.

Location is another important factor that can affect the amount of crop loss due to weeds. In some parts of the world, certain types of weeds may be more prevalent and cause higher levels of crop loss. For example, in tropical regions, some weed species may be more adapted to the warm, humid conditions and may be more competitive with crops.

To reduce crop losses due to weeds, farmers can use a variety of weed management techniques. These can include mechanical control methods, such as tilling the soil to disrupt the growth of weeds, or chemical control methods, such as applying herbicides to kill weeds. Cultural control methods, such as crop rotation and cover cropping, can also be effective in reducing weed pressure and reducing crop losses due to weed competition.

Overall, weeds are a major cause of crop losses on farms around the world, and they can significantly impact a farmer's livelihood and the productivity of a farm. By using effective weed management techniques, farmers can help to reduce crop losses due to weeds and improve the overall yield of their crops.

There are several reasons why farmers need to control weeds:

- **Weeds can reduce crop yields and quality.** Some are known to have allelopathic chemicals, which can inhibit the growth of nearby plants. These chemicals can also affect the quality of crops, leading to reduced market value. Invasive plants can also physically compete with crops for space, light, water, and nutrients, leading to reduced crop growth and yields.
- **Weeds can increase production costs.** Controlling invasive plants can be time-consuming and labor-intensive, requiring frequent monitoring and management. Farmers may need to use herbicides, mechanical controls, or other methods to control the spread of invasive plants, which can increase production costs.

DOI: 10.1079/9781836992288.0001

- **Weeds can cause damage to farm equipment.** Some invasive plants have sharp thorns or burrs that can damage farm equipment and machinery, leading to costly repairs and downtime.
- **Weeds can reduce the value of agricultural land.** Invasive plants can alter the natural ecosystem, leading to a decrease in the value of agricultural land. This can have a negative impact on the value of the farm and the profitability of the business.

Other than negative effects on agriculture, weeds can have a number of negative effects on urban infrastructure. In urban areas, weeds can grow in and around buildings, sidewalks, and roads, causing aesthetic and maintenance problems. Weeds can affect the integrity of infrastructure by growing into or around foundations, walls, and other structures. This can cause damage to the infrastructure and make it more prone to collapse or failure. Weeds can also be a safety hazard by growing over or obscuring road signs, traffic signals, and other important indicators. This can cause confusion and increase the risk of accidents. In addition, weeds can harbor pests and diseases that can spread to other plants, causing further damage. Weeds can also serve as a breeding ground for insects and other pests that can be harmful to humans, such as mosquitoes and ticks. Weeds can also contribute to soil erosion and degradation, which can have long-term impacts on the environment. This can be particularly problematic in urban areas where the soil is often compacted and the vegetation is limited, leading to increased erosion and sedimentation.

Overall, it is important to manage weeds in and around urban infrastructure in order to minimize their negative effects and maintain the integrity, safety, and aesthetics of the infrastructure. This can be done through a combination of chemical and nonchemical control methods, such as physical removal, mulching, and the use of herbicides.

In conclusion, it is essential for farmers to implement effective management strategies to prevent the spread of invasive plants and protect their agricultural operations, and to control the potential damage to urban infrastructure caused by weeds.

Herbicides

An herbicide is a chemical substance that is used to kill or control the growth of plants, specifically unwanted plants or weeds. Herbicides are often used in agriculture, horticulture, and landscaping to control the growth of weeds and other undesired plants, as they can compete with crops or ornamental plants for space, nutrients, and water. They can be applied to the soil, leaves, or roots of plants and are typically selective, meaning that they are formulated to kill certain types of plants while not harming others. Some herbicides are also nonselective, meaning that they will kill any plant they come into contact with. It is important to carefully follow the instructions on the label when using herbicides, as they can be toxic to humans and other animals.

It is to be considered that the use of herbicides can have both positive and negative impacts on society. On the positive side, herbicides can help to control the spread of invasive or nuisance plants, which can have negative impacts on native ecosystems and agriculture. Herbicides can also be used to manage weeds in urban areas, which can improve the appearance and safety of public spaces. However, the use of herbicides also has the potential to have negative impacts on the environment and human health. Some herbicides can be toxic to nontarget plants and animals, and can have negative impacts on soil health. There is also evidence that certain herbicides may have negative impacts on human health, including potential links to cancer and other health problems. In addition, the use of herbicides can also have economic costs, as they may be expensive to purchase and apply, and may require specialized equipment. There may also be costs associated with the disposal of herbicides and their containers.

Given the potential negative impacts of herbicide usage, it is important to carefully consider the costs and benefits of using herbicides and to use them in a responsible and sustainable manner. This may involve applying herbicides only when necessary, using the minimum amount needed to achieve the desired effect. Some of the potential environmental costs of herbicide usage include:

- **The contamination of water resources.** Some herbicides can leach into the soil and eventually make their way into surface or ground water, where they can be harmful to aquatic life and potentially contaminate drinking water sources.
- **Damage to nontarget plants.** Herbicides can sometimes harm or kill plants that were not intended to be targeted, which can have negative impacts on ecosystems and wildlife that depend on those plants.
- **Damage to soil.** Some herbicides can harm or kill beneficial microbes in the soil, which can lead to reduced soil health and fertility.
- **Pollution.** The production and use of herbicides can result in the release of pollutants into the air and water, which can have negative impacts on the environment and human health.

Overall, the simple fact is that synthetic herbicides are widely used, representing over 40% of the world's agrochemical market (Grube *et al.*, 2004).

Glyphosate

The story of the development of glyphosate, one of the world's most widely used herbicides, is quite fascinating. Glyphosate was invented in 1950 by a Swiss chemist, Dr Henri Martin, who worked for the small pharmaceutical company, Cilag (Franz *et al.*, 1997), but it was not reported as an herbicide at the time. As an herbicide, glyphosate was first discovered and developed by a chemist named Dr John E. Franz while working at the Monsanto Company in the early 1970s.

In the late 1960s, under Dr Phil Hamm, Monsanto was researching chemicals that could be used as potential water-softening agents. During this research, Dr Franz synthesized a series of compounds that included glyphosate. However, it was initially not recognized for its herbicidal properties. In 1970, while screening the synthesized compounds for their potential applications, Dr Franz noticed that glyphosate had remarkable weed-killing properties. Glyphosate exhibited broad-spectrum herbicidal activity, meaning it could effectively control a wide range of weeds without significant harm to most crops. Recognizing the potential of glyphosate as an herbicide, Dr Franz and his team at Monsanto began extensive testing and development efforts. They discovered that glyphosate worked by inhibiting an enzyme called 5-enolpyruvylshikimate-3-phophate (EPSP) synthase, which is crucial for the synthesis of certain amino acids in plants. This inhibition disrupted the plant's metabolic processes, leading to its death.

Monsanto filed a patent application for glyphosate in 1971, and the herbicide was first introduced to the market in 1974 under the brand name Roundup. Roundup quickly gained popularity due to its effectiveness, low toxicity to animals and humans, and its ability to degrade in the environment (Dill *et al.*, 2020).

As the use of glyphosate increased, Monsanto faced challenges in obtaining regulatory approvals for its herbicide. While the widespread use of glyphosate has made it an effective tool for controlling weeds, it has also been the subject of controversy and debate. Some studies have suggested that the herbicide may have negative effects on human health and the environment, while others have found no such effects. As a result, the use of glyphosate and products containing it, such as Roundup, has been the subject of regulatory scrutiny in some countries. It has been classified as a possible human carcinogen by the International Agency for Research on Cancer (IARC).

According to Duke and Powles (2008), glyphosate works by inhibiting the enzyme EPSP synthase, which is critical for the synthesis of essential amino acids in plants, leading to their death. Its effectiveness at relatively low doses and its unique mode of action made it a revolutionary "once-in-a-century herbicide." However, the widespread use of glyphosate has led to the evolution of resistant weeds, requiring higher application rates or alternative weed management strategies.

Therefore, even with the regulatory challenges, potential environmental and health-related risks, growing weed resistance, and operational hurdles, such as drifting and lixiviation, glyphosate is the most commonly used herbicide by far, being a cheap broad-spectrum systemic herbicide. These facts speak to the potential of a technology that may substitute for it.

Genetically modified organism seeds

Genetically modified organisms (GMOs) are living organisms whose genetic material has been artificially modified in a laboratory using genetic engineering techniques. This is done in order to introduce new traits or characteristics into the organism that would not naturally occur through traditional breeding methods.

The development of GMO seeds is a complex and multifaceted story that spans several decades. Here is a brief overview of the key milestones and events in the development of GMO seeds.

- **Early research.** The groundwork for GMO seeds was laid in the 1970s when scientists started exploring genetic engineering techniques. They discovered ways to manipulate the genetic material of plants by introducing specific genes from other organisms to confer desired traits.
- **First GM crop.** The first commercially available GMO crop was the Flavr Savr tomato, developed by the company Calgene in the early 1990s. It was genetically engineered to delay ripening and improve shelf life.
- **Herbicide-tolerant crops.** In the mid-1990s, Monsanto introduced Roundup Ready crops, which were genetically modified to be tolerant to glyphosate, the active ingredient in the herbicide Roundup. These crops allowed farmers to spray glyphosate to control weeds without harming their crops.
- **Insect-resistant crops.** Around the same time, scientists developed crops engineered with *Bacillus thuringiensis* (Bt) genes. These genes produce proteins toxic to certain insect pests, reducing the need for synthetic insecticides. Bt cotton and Bt corn were among the first insect-resistant GM crops widely adopted by farmers (Sanahuja *et al.*, 2011).
- **Expansion of GMO crops:** The adoption of GMO seeds expanded rapidly in the late 1990s and early 2000s. Companies, such as Monsanto, Syngenta, DuPont, and Bayer CropScience, developed a variety of GMO crops, including soybeans, canola, alfalfa, sugar beets, and more. These crops were engineered for traits such as herbicide tolerance, insect resistance, disease resistance, and improved yield.
- **Regulatory oversight.** The development and commercialization of GMO seeds raised concerns about their safety and potential environmental impacts. Governments and regulatory bodies worldwide implemented systems to assess and regulate GMO crops. Different countries have varying approaches to GMO regulations, with some imposing strict restrictions or bans, and others allowing cultivation and importation with specific guidelines.
- **Public debate and controversies.** GMO seeds and crops sparked public debates and controversies, particularly regarding food safety, environmental impacts, and corporate control of the seed supply. Advocacy groups, scientists, policymakers, and consumers have expressed differing viewpoints on the benefits and risks associated with GMOs.
- **Advancements in trait stacking and genomic editing.** In recent years, advancements in genetic engineering techniques have allowed for more precise modifications of plant genomes. Trait stacking involves combining multiple desirable traits in a single plant, such as insect resistance, herbicide tolerance, and improved yield. Additionally, newer technologies, such as clustered regularly interspaced short palindromic repeats (CRISPR)-Cas9, have enabled targeted genomic editing, providing opportunities for crop improvement without introducing genes from unrelated species.

The development of GMO seeds is an ongoing process with continued research, innovation, and debates surrounding their use and regulation. It is worth noting that regulations and public opinion regarding GMOs may vary across different countries and regions.

Those seeds usually are protected by intellectual property (IP) regulations, such as patents, so farmers cannot use the production of one cycle as seeds for the next. This means a significant cost increase in the form of royalties to be paid to agrochemical companies. Therefore, herbicide-resistant GMO seeds, particularly those engineered to tolerate glyphosate (e.g. Roundup Ready crops), have both advantages and disadvantages.

On the positive side, these seeds offer enhanced weed control for farmers. They can apply specific herbicides without causing harm to their crops, leading to improved weed management and potentially higher yields. This convenience and efficiency can save farmers time and resources, as they can reduce labor-intensive manual weeding or the need for multiple herbicide applications. Another potential benefit is the reduction in pesticide use. Herbicide-resistant GMO seeds can reduce the reliance on synthetic herbicides and other chemical weed control methods. This reduction may have positive environmental implications, such as decreased chemical runoff into water sources.

However, there are also concerns associated with herbicide-resistant GMO seeds. One major drawback is the emergence of herbicide-resistant weeds over time. Prolonged use of the same herbicide can lead to the adaptation and development of resistance in certain weed species. This can result in increased reliance on alternative herbicides or the need for additional weed management practices, reducing the effectiveness of herbicide-resistant crops. The financial and operational costs associated with herbicide-resistant weeds can vary depending on factors such as the specific crop, region, farming practices, and the level of herbicide resistance. Herbicide-resistant crops have been expanding in the farming market, but their cost implications are important to consider. One significant aspect is the cost of herbicide-resistant seeds, which can be higher than conventional seeds. This price difference is due to the investment in research and development, IP rights, and the benefits these seeds offer to farmers. However, the actual cost varies among crop varieties and seed suppliers.

In addition to seed costs, farmers need to consider the expenses associated with herbicides. While herbicide-resistant crops provide targeted weed control, the cost of herbicide products and their application should be taken into account. This includes the purchase price of the herbicides, as well as any additional costs related to equipment, labor, and training for their proper and safe use.

Proper resistance management is crucial for maintaining the effectiveness of herbicide-resistant crops. Farmers must implement strategies such as integrated weed management, rotation of herbicides or modes of action, and cultural practices to minimize weed pressure. These practices may result in additional costs, such as purchasing a variety of herbicides and potentially requiring more labor-intensive weed management techniques. Moreover, farmers may need to invest in specialized equipment and infrastructure to effectively utilize herbicide-resistant crops. This can include sprayers, applicators, or precision agriculture technologies, which can contribute to the operational costs associated with these crops.

While herbicide-resistant crops have been expanding in the farming market, it is important to assess the costs and benefits on a case-by-case basis. Farmers evaluate factors such as crop profitability, weed pressure, local regulations, and their own management capabilities to determine the feasibility and financial implications of adopting herbicide-resistant crops in their operations.

Furthermore, the widespread use of herbicide-resistant GMO seeds may impact biodiversity (Jacobsen *et al.*, 2013). The dominance of herbicide-tolerant crops can reduce the variety of plant species in agricultural ecosystems, potentially affecting natural habitats and beneficial insect populations.

It is important to note that the impact and perception of herbicide-resistant GMO seeds can vary among different stakeholders, and ongoing research and monitoring are essential to address potential challenges and maximize the benefits of these seeds.

In other words, as herbicide-resistant GMOs are indeed a great tool for farming, that allow for an easier operation with deflationary effects on the world's food economy, they also have social, economic, and environmental costs, and other negative implications, such as the dependency of farmers to the will of chemical companies with potentially misaligned interests.

Resistance to herbicides

Resistance to herbicides is the consequence of the overuse that can lead to the development of herbicide-resistant weeds, which can be more difficult and costly to control. This can, in turn, lead to the need for even more herbicide molecules, mixes, and volume to be used. In the ongoing debate surrounding GMO crops, one of the most contentious issues is the rise of herbicide-resistant weeds, commonly known as superweeds. These weeds have evolved to withstand glyphosate, the widely used herbicide marketed as Roundup by Monsanto. Initially, glyphosate-resistant crops allowed farmers to eliminate weeds effectively without harming their crops. However, overreliance on this single herbicide has accelerated the natural selection process, leading to the proliferation of superweeds.

These herbicide-resistant superweeds threaten agricultural productivity. Palmer amaranth, a particularly aggressive species, has rapidly spread across U.S. farmland, outcompeting crops for resources. The overreliance on glyphosate has accelerated weed adaptation, forcing farmers to rethink their approach to weed management (Gilbert, 2013). In response, many farmers have increased herbicide application rates or reverted to mechanical tilling, both of which come with environmental and economic drawbacks. Monsanto initially dismissed concerns about resistance, but, by 2011, glyphosate-resistant weeds had spread across 76 counties in the U.S. state of Georgia alone.

There are currently 534 unique cases (species × site of action) of herbicide-resistant weeds globally, from 273 species (156 dicots and 117 monocots) (Fig. 1.1). Weeds have evolved resistance to 21 of the 31 known herbicide sites of action and to 168 different herbicides. Herbicide-resistant weeds have been reported in 102 crops in 75 countries (Fig. 1.2; Heap, 2018). In other words, "superweeds" have become a global issue, affecting key agricultural regions in Brazil, Argentina, and Australia.

To counteract resistance, biotechnology firms have developed crops resistant to alternative herbicides such as dicamba and 2,4-D. However, experts warn that continued reliance on chemical solutions may only perpetuate the cycle of resistance. Instead, integrated weed management—combining chemical, mechanical, and biological methods—offers a more sustainable path forward. By rotating herbicides, planting cover crops, and employing targeted tillage, farmers can slow the evolution of resistant weeds. These diversified strategies reduce reliance on single herbicide solutions and promote long-term agricultural sustainability.

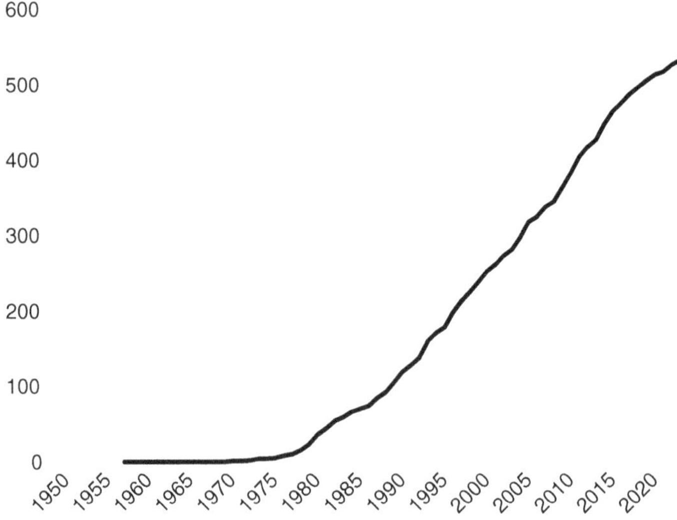

Fig. 1.1. Number of unique cases of herbicide resistance (Heap, 2018). (Figure used with permission from Dr Ian Heap.)

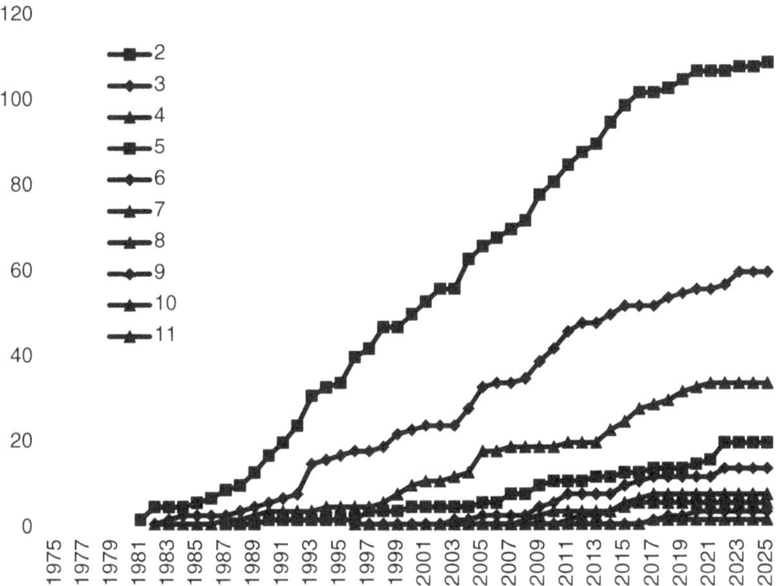

Fig. 1.2. Number of weeds resistant to herbicide modes of action (Heap, 2018). (Figure used with permission from Dr Ian Heap.)

The rise of superweeds highlights the challenges of industrial farming's dependence on chemical solutions. While GM crops have increased yields, they have also created new ecological problems. This poses a great operational challenge for the continued use of herbicides and creates an intrinsic demand for the development of new weeding methods. Moreover, there are also multiple resistance cases, where some weeds, sometimes denominated "superweeds," develop resistance to multiple modes of action, as shown in Fig. 1.1 and Fig. 1.2:

Therefore, substituting herbicides is not an "if" question, although "when" and "how" may not be yet accomplished. In a paper regarding the best fit for electric weed control (EWC)in Australia, the authors Borger and Slaven (2023) state that "*A major reason for utilizing alternative weed control tactics is the management of herbicide resistant weeds. For example, weed control along fence lines may have limited herbicide options as growers want to protect non-target native species. Overreliance of non-selective herbicides in these areas commonly leads to resistance. Electric weed control is an alternative technology to control these resistant weeds. For example, Borger and Slaven (2023) applied electric weed control to a population of annual ryegrass that was resistant to 4 L ha^{-1} glyphosate (Roundup PowerMax®). Electric weed control at 2–4 km h^{-1} provided control comparable to an application of glyphosate 855 g a.i. ha^{-1} (Roundup UltraMax® 570 g a.i. L^{-1}, Bayer, SC) followed by paraquat/diquat 202/172 g a.i. ha^{-1} (Spray.Seed® 135/115 g a.i. L^{-1}, Syngenta, SL) (i.e. a herbicide 'double knock').*"

Control of varying weed species in trials was conducted in Western Australia in 2022 and 2023 (Borger and Slaven, 2023). All weed species in Table 1.1 were mature at the time of EWC application, except for annual ryegrass where both young and mature plants were treated.

Figure 1.3 shows the effect of a single or double application of EWC at 2 or 4 km h^{-1} on the dry biomass of an annual ryegrass population at a trial site at the Department of Primary Industries and Regional Development, Northam, Western Australia, in 2023. Horizontal bars indicate the standard error of eight replications. Bars annotated with the same letter have means that were not significantly different ($P<0.001$, LSD: 116.0). The data are recreated from (Borger and Slaven, 2023).

Table 1.1. Control of varying weed species in trials conducted in Western Australia in 2022 and 2023 (Borger and Slaven, 2023). (Table used with permission from Western Australian Agricultural Authority.)

Weed species	Control
Annual ryegrass (*Lolium rigidum* Gaud.)	Over 90% at 1.5 km h^{-1}, 40–95% at 3 km h^{-1}. Easy to kill young plants. However, results are variable for mature plants.
Kikuyu (*Cenchrus clandestinus* Hochst. ex Chiov.)	Over 95% at 1.5 and 3 km h^{-1}.
Wild radish (*Raphanus raphanistrum* L.)	Over 80% at 2–4 km h^{-1}. Mature plants may resprout.
Capeweed (*Arctotheca calendula* (L.) K.Lewin)	Over 85% at 2–4 km h^{-1}.
Erodium (*Erodium* spp.)	Over 85% at 2–4 km h^{-1}.
Couch grass (*Elymus repens* (L.) Gould)	Does not give full control of mature plants at 2 km h^{-1}.
Guildford grass (*Romulea rosea* (L.) Eckl.)	Does not give full control of mature plants at 2 km h^{-1}.
Winter grass (*Poa annua* L.)	Does not give full control of mature plants at 2 km h^{-1}.
Soursob (*Oxalis pes-caprae* L.)	Over 85% at 2–4 km h^{-1}.
Medic (*Medicago* spp.)	Over 85% at 2–4 km h^{-1}.
Cape tulip (*Moraea flaccida* (Sweet) Steud.)	All above ground vegetation removed at 2–4 km h^{-1}. Bulb survival/regrowth was lower following electric weed control at 2 km h^{-1} compared to herbicide (glyphosate 570 g a.i. l^{-1}, followed by paraquat/diquat 135/115 g a.i. l^{-1}).
Windmill grass (*Chloris truncata* R.Br.)	Partial control (35–90%) with double application treatments at 2–4 km h^{-1}. A clump of windmill grass includes multiple plants and some plants are protected by neighbors.
Wireweed (*Polygonum aviculare* L.)	Over 95% at 2–4 km h^{-1}.
Sowthistle (*Sonchus oleraceus* L.)	Over 95% at 2–4 km h^{-1}.
Flatweed (*Hypochaeris radicata* L.)	60–90% at 2–4 km h^{-1}.
Annual veldt grass (*Ehrharta longiflora* Sm.)	Over 90% at 1.4 km h^{-1}.
Silver grass (*Vulpia myuros* (L.) C.C. Gmel.)	Over 90% at 1.4 km h^{-1}.
Common dandelion (*Taraxacum officinale* L.)	Over 90% at 1.4 km h^{-1}.

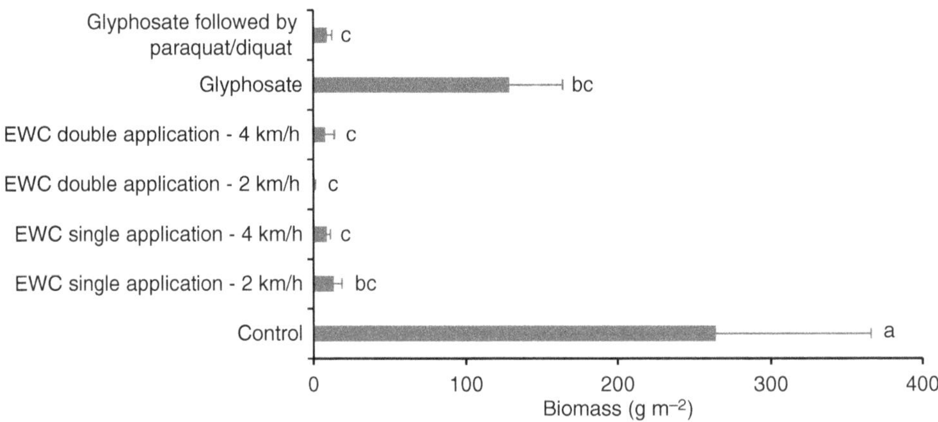

Fig. 1.3. Case of resistant weed controlled by electric weed control (EWC) in comparison with herbicides. (Figure used with permission from Slaven and Borger.)

References

Borger, C.P.D. and Slaven, M.J. (2023) *What Is the Best Fit for Electric Weed Control in Australia?* Australian Department of Primary Industries and Regional Development, Perth, Western Australia.

Dill, G.M., Sammons, R.D., Feng, P.C.C., Kohn, F. and Kretzmer, K. (2020) *Glyphosate: Discovery, Development, Applications, and Properties*. John Wiley & Sons, Hoboken, NJ, USA.

Duke, S.O. and Powles, S.B. (2008) Glyphosate: aonce-in-a-century herbicide. *Pest Management Science* 64(4): 319–325.

Franz, J.E., Mao, M.K. and Sikorski, J.A. (1997). *Glyphosate: A Unique Global Herbicide.* American Chemical Society, ACS Monograph 189., Washington, DC, USA.

Gilbert, N. (2013) Case Studies: a hard look at GM crops. *Nature* 497: 24–26.

Grube, A.H., Donaldson, D. and Kiely, T. (2004) *Pesticides Industry Sales and Usage: 2000 and 2001 Market Estimates*. U.S. Environmental Protection Agency, Washington, DC, USA.

Heap, I. (2018) The International Herbicide-Resistant Weed Database. Available at: https://www.weedscience.org/ (accessed Octobe 27, 2025).

Jacobsen, S.E., Sørensen, M., Pedersen, S.M. and Weiner, J. (2013). Feeding the world: genetically modified crops versus agricultural biodiversity. *Agronomy for Sustainable Development* 33: 651–662.

Sanahuja, G., Banakar, R., Twyman, R.M., Capell, T. and Christou, P. (2011) *Bacillus thuringiensis: A Century of Research, Development and Commercial Applications*. Wiley, Oxford, UK.

2

The History of Electrical Weeding

The Technological Journey of Electrical Weeding

The journey toward developing sustainable, nonchemical weed management methods has been marked by a fascinating evolution of technology, dating back to the late 19th century. In the face of ongoing agricultural challenges, innovators have sought alternatives to labor-intensive manual weeding and the environmentally harmful use of chemical herbicides. From early electromechanical devices to today's advanced, high-frequency systems, this chapter delves into the history and technological advances in electric weed control (EWC), examining how each step in this evolution has laid the groundwork for modern, sustainable solutions. At the core of these innovations is the idea of using electricity to disrupt weed growth, presenting a promising approach that is efficient, precise, and environmentally sound. By tracing this development, we can understand how each invention has built upon previous efforts, creating a robust foundation for today's cutting-edge systems.

One of the earliest recorded efforts in EWC dates back to Albert A. Sharp's invention of the "Vegetation Exterminator" in 1893. This pioneering concept sought to disable weeds using high-voltage pulses applied directly to the plants. Sharp's work marked the beginning of an era of experimentation, laying the conceptual groundwork for electrical weed management. Despite its promise, however, Sharp's early prototype and similar devices faced significant limitations. The technology of the time was inadequate to deliver controlled, sustained power safely or practically in open agricultural settings. Additionally, these early devices often lacked precision, making it difficult to target weeds without harming nearby crops. Nonetheless, Sharp's invention captured the agricultural sector's imagination and inspired subsequent generations of inventors to explore the potential of electricity as a nonchemical means of weed control.

As the 20th century progressed, advances in electrical engineering provided new tools for inventors seeking to refine Sharp's concept. Developments in transformer technology, frequency modulation, and power control systems made it possible to design more compact and efficient devices. During this period, researchers and agricultural engineers sought ways to better manage and distribute high-voltage currents, minimizing the risk of uncontrolled arcs and improving the accuracy of EWC methods. While progress was made, these early devices were often heavy, costly, and required significant manual operation, limiting their practicality for widespread agricultural use. Nonetheless, the groundwork laid during this period contributed valuable insights into the application of

DOI: 10.1079/9781836992288.0002

high-voltage pulses for selective weed control and paved the way for the next phase of innovation.

The late 20th and early 21st centuries witnessed a surge in technological advancements that would prove transformative for EWC. Innovations in electronics, including solid-state devices, high-frequency inverters, and capacitive voltage multipliers, enabled more precise and reliable control of high-voltage systems. These new components allowed inventors to design systems that could adapt to varying weed densities and types, significantly enhancing the flexibility and efficiency of electric weed management devices. The advent of modular designs, frequency converters, and harmonic filters provided better control over the energy applied to weeds, allowing for selective and efficient treatment. By addressing some of the key limitations of earlier systems, these advancements brought EWC closer to practical reality, offering a viable, nonchemical alternative to herbicides in certain agricultural applications.

Today, Zasso's EWC technology represents the culmination of more than a century of incremental innovation. Building on the principles established by Sharp and subsequent inventors, Zasso has developed high-frequency weed control systems that are both effective and sustainable. Zasso's technology incorporates a modular design, making it adaptable to different types of agricultural settings and scales. High-frequency inverters and capacitive voltage multipliers enable precise voltage control, allowing the device to deliver just the right amount of energy for each weed situation, where, with the idea of the impedance matcher, the state of the art goes back to direct current (DC) output. The result is a solution that is both efficient and environmentally friendly, eliminating the need for harmful chemicals while protecting nearby crops and soil health. Furthermore, Zasso's systems include advanced impedance-matching properties, which not only prevent resonance issues but also enhance energy efficiency, contributing to the durability and reliability of the equipment.

This chapter will explore the progression from early concepts to Zasso's state-of-the-art systems, examining each invention's unique contributions to the field of EWC. Through an in-depth analysis of these technologies, readers will gain insight into the engineering challenges and breakthroughs that have defined this journey. From the limitations of early prototypes to the sophisticated, high-frequency devices available today, the evolution of EWC reflects the broader technological advancements in agriculture and environmental stewardship. As chemical herbicides face increasing scrutiny due to their environmental and health impacts, EWC emerges as a promising alternative that aligns with modern sustainability goals. By understanding the history and innovations that brought us here, we can appreciate the transformative potential of these technologies for the future of agriculture.

The "Vegetation Exterminator"

In 1893, Albert A. Sharp's invention, the "Vegetation Exterminator," was the first device that was designed to efficiently clear unwanted plant growth from railway beds, highways, and fields (Fig. 2.1). In regions with warm climates, vegetation grows rapidly along roads and railways, often obstructing paths and requiring regular clearing. This invention offered a straightforward solution by using electric currents to eliminate unwanted vegetation as the vehicle moves along the track or road (Sharp, 1893).

The Vegetation Exterminator was a mobile apparatus that could be mounted on any vehicle, though it is illustrated here on a push car running along railroad tracks. Its primary component was a metallic brush connected to an electric power source, which applied high-voltage currents directly to the vegetation, killing it upon contact. This electric brush was particularly effective for thick roadside growth and could also be adapted to destroy weeds in cultivated fields without harming desired crops.

The apparatus includes several key components:

- **Vehicle setup.** The exterminator can be mounted on various vehicles, with this example showing it mounted on a push car. The push car has a housing to protect the dynamo and other electric components.
- **Power source.** An electric dynamo or engine generated the electric current required to kill vegetation. In this example,

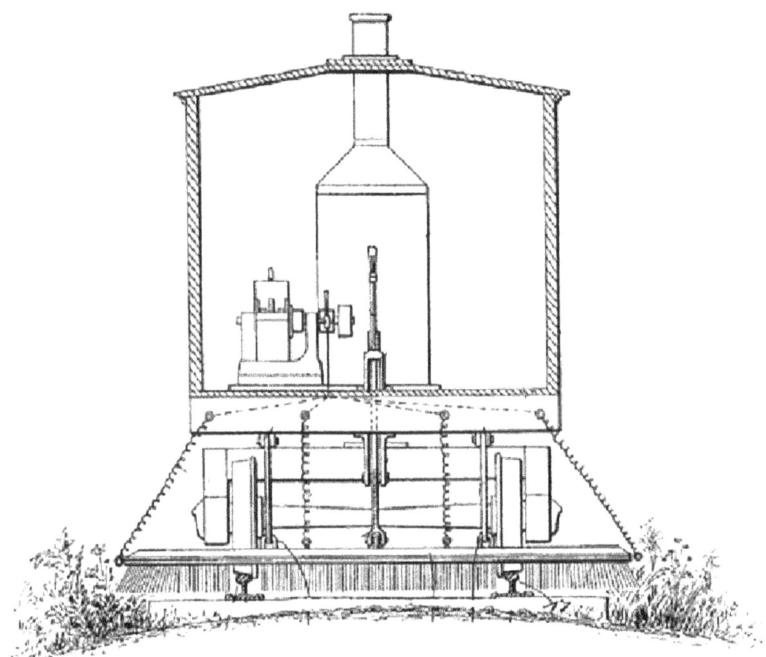

Fig. 2.1. The "Vegetation Exterminator." (Sharp, 1893).

the dynamo is engine-driven, though alternative power sources such as batteries could be used.

- **Metallic brush.** Suspended below the vehicle and spanning the width of the track or road, the brush contained metal wires that made direct contact with plants. The brush was configured to avoid the tracks by leaving gaps in the wires where it straddled the rails. Connected to the dynamo by electric wires, the brush conducted current through the plants it contacted, grounding the current to destroy the plants effectively.
- **Adjustable lever mechanism.** A lever system allowed the brush to be raised, as necessary, when approaching obstructions such as bridges or guard rails. The lever, which was mounted within easy reach, could lift the brush to avoid damage or clear larger obstacles.

As the vehicle moved along the track, the dynamo generated electricity, sending current through the metallic brush, which contacted and destroyed vegetation along the way. The lever allowed for easy adjustment of the brush's position, ensuring flexibility across different terrains and obstacles. This design enabled continuous, low-maintenance vegetation control, which was ideal for railways and highways, as well as large-scale agricultural fields.

The Vegetation Exterminator represented the first attempt at a practical approach to vegetation management by integrating mechanical and electrical elements to achieve a streamlined, effective process. Its use in road maintenance could reduce the reliance on manual labor and herbicides, offering a more sustainable method for maintaining clear paths and promoting efficient agricultural operations.

The "Weed Destroyer"

Several years later, similar concepts of electrical applications for vegetation control followed and innovations were registered (Burt, 1928). In 1928, W.E. Burt proposed an arrangement for producing a powerful current that was mounted on a mobile platform, designed to be easily transported from one location to another (Fig. 2.2). This setup comprised a motor, a DC generator (or exciter), and an alternating current (AC) generator. These generators were driven by belts from the motor, and a switchboard was mounted on springs attached to a supporting frame on the truck. The exciter was electrically connected with the field of the AC generator through a switch, and the system included essential monitoring and control instruments such as voltmeters, ammeters, and rheostats. Additionally, the secondary circuit was arranged with a transformer, an adjustable gap, condensers, and an inductance coil, configured to produce a more powerful current than previously available methods, although the basic operational approach remained the same.

Various arrangements for producing electric energy could be utilized to execute this method of eradicating unwanted plants, though an oscillatory current is preferable. From this description, the advantages and innovative aspects of this invention were expected to be clear. However, it is important to note that modifications to the construction or adjustments to the combination and arrangement of parts were within the scope of what the invention claimed.

The invention also specified the means for exterminating plants with electricity, including an AC generator, a DC exciter generator connected to it, a motor for operating both generators, and a transformer connected to the AC generator. This transformer was linked to a condenser and then to a cutter, with an essential spark gap in the connection and a ground rod electrically connected with the condenser. Another version of this arrangement incorporated a truck to support the various components, making it a practical and transportable solution for plant eradication.

The "Electrical Weed Killer"

In 1947, Gilbert M. Baker's invention, the "Electrical Weed Killer" (US2682729A), introduced an innovative approach to weed control by using high-voltage electricity to destroy weeds. Mounted on a vehicle, this electrical weed killer applied a powerful current directly to exposed portions of weeds, ensuring that electricity flowed through the entire plant structure, including the roots (Fig. 2.3). Unlike traditional methods, this device didn't rely on moisture in the soil to transmit current effectively, making it suitable for various conditions (Poynor, 1954).

Historically, EWC has involved driving current through the ground. Typically, this was done by embedding metal stakes or connecting to grounded structures like iron rails, then positioning an electrode to contact the upper portions of the weeds. This setup forced current to travel through the soil and up into the weeds. However, the method posed limitations: if the ground was damp, the electric current would often burn the visible parts of the weed but leave the roots largely

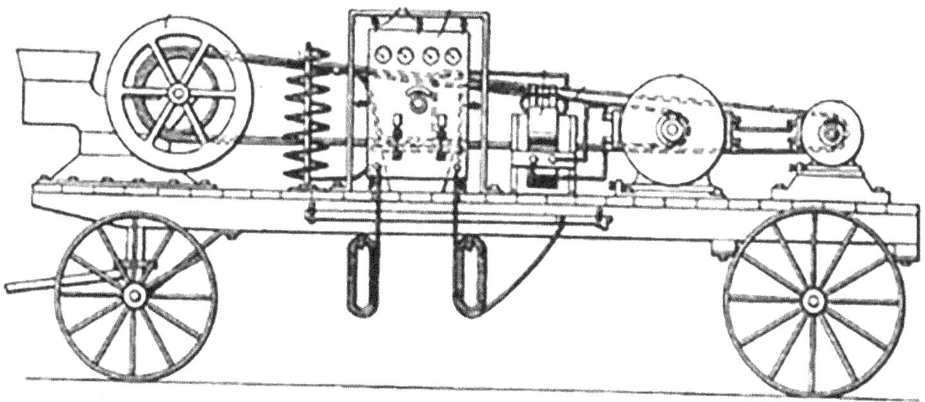

Fig. 2.2. The "Weed Destroyer." (Burt, 1928).

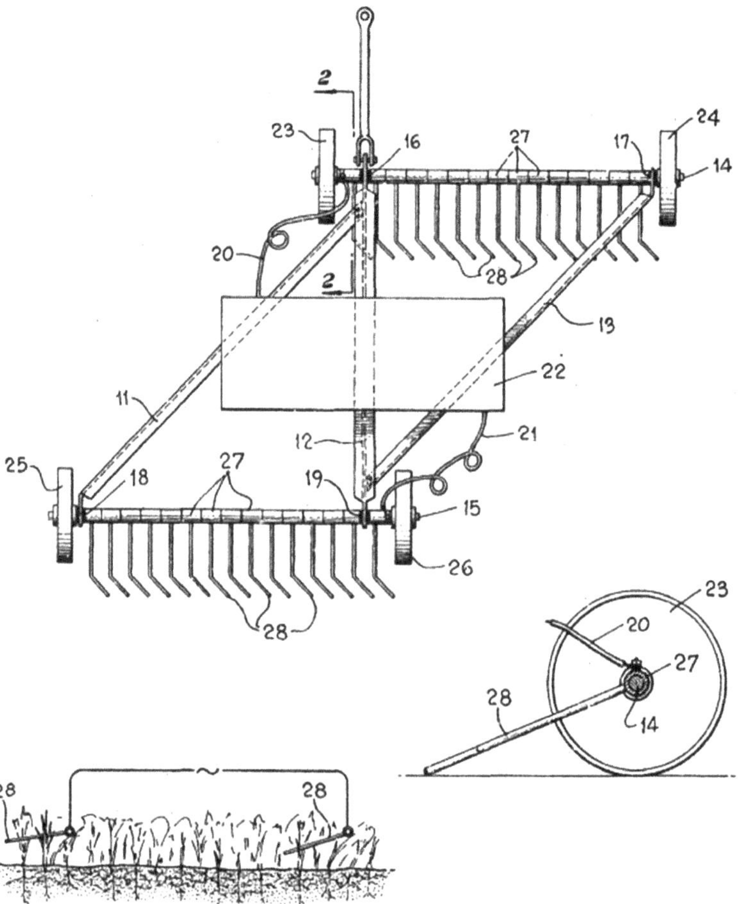

Fig. 2.3. The "Electrical Weed Killer."

unharmed. Moving stakes around each treated area also proved cumbersome and inefficient.

The electrical weed killer design solved these problems by using two electrodes mounted parallel to each other along the underside of a vehicle. These electrodes made direct contact with the ground and the weeds. As the vehicle moved, the electrodes brushed across weeds, connecting with exposed surfaces and allowing electricity to travel continuously from one set of plants to the next. This flow forced current through the weeds' root systems, effectively destroying each plant from the inside out, regardless of soil moisture.

This apparatus was powered by a dynamo that generated an AC, which was intensified by a transformer to approximately 12,500 V. This high-voltage supply was distributed evenly across the electrodes. The vehicle, insulated from its wheels and frame, used sleeves and contact rods to maintain stable contact with weeds of different heights. These rods were designed to rise and fall over uneven ground, ensuring constant contact for a thorough treatment.

As shown in the simplified schematic, the device relied on two main components: the parallel electrode system (Fig. 2.3, no. 27) and the high-voltage power source. This configuration ensured that current moved evenly across treated weeds, minimizing the need for exact ground moisture levels and guaranteeing comprehensive plant destruction. The design also allowed for easy modification and adaptability, such as using chain loops as electrodes in place of the standard rods.

By connecting high voltage directly to each electrode, the invention addressed earlier limitations in weed control technology. It ensured uniform current flow through all weeds in the device's path, providing a reliable and efficient solution for large-scale weed management.

The "Electric Row-Crop Thinning Machine"

In agricultural operations, young crops are often planted densely to ensure enough viable plants survive for a full, healthy harvest. However, once established, these crops require thinning to allow individual plants the space to grow. In 1953, Noal L. McCreight and John H. McCreight's patented Electric Row-Crop Thinning Machine was designed specifically for this purpose (Fig. 2.4). The machine thins crops by removing unwanted plants along a row, using high-voltage electricity to destroy targeted plants while leaving others untouched (McCreight and McCreight, 1953).

The central idea of this invention was a specialized plant-contact assembly that enabled the machine to remove plants at predetermined intervals along a crop row. By spacing out these "blocking" points, the machine selectively eliminated plants to provide enough room for the remaining ones to flourish. The thinning process was achieved by applying high-voltage electricity through unique, rotating contact rollers that briefly engaged the plants at each preset interval.

The machine itself was built as a trailer and could be connected to any suitable motorized vehicle. The trailer was equipped with insulated cross beams, which supported the contact rollers and kept the electrical components isolated from the rest of the machine. A generator and transformer inside the trailer converted power from the towing vehicle into a high-voltage charge used to kill the unwanted plants.

Each contact roller was engineered to perform selective thinning. The rollers featured

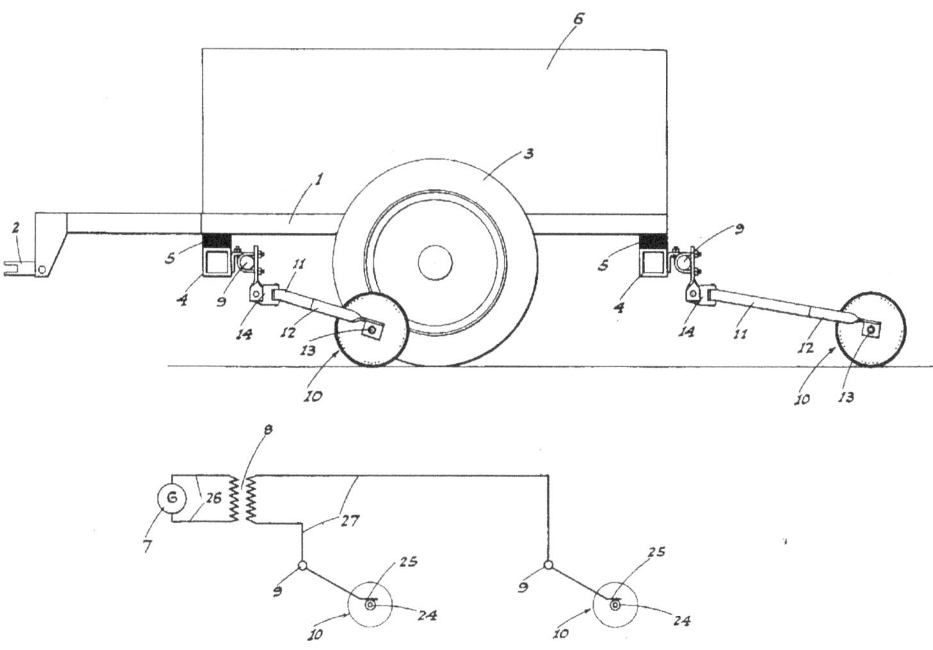

Fig. 2.4. The "Electric Row-Crop Thinning Machine." (McCreight and McCreight, 1953).

evenly spaced cross rods that extended between two dielectric side discs, spaced circumferentially around each roller. These rods were strategically spaced to engage plants at intervals, delivering a brief, high-voltage pulse that effectively destroyed the plants on contact. As the machine moved forward, only those plants that came into direct contact with the charged rods were eliminated, allowing other plants to remain untouched and continue growing.

The electrical circuit connecting the power source to the rollers ensured an efficient and controlled energy flow. The high-voltage circuit was grounded between each roller and spaced sufficiently to prevent any unintended arcing. As the machine advanced along a crop row, the rotating rollers made contact with plants at spaced intervals, applying high-voltage electricity directly to each targeted plant and eliminating it instantly.

The roller assembly included dielectric side discs to prevent unwanted conductivity and rubber tires to absorb impact and reduce wear. In some models, a commutator ring was connected to the roller, intermittently charging the roller to apply electricity only when specific points on the crop row were reached. This design conserved energy and ensured precise spacing for the crop-thinning process.

This machine was designed with practicality in mind. The high-voltage thinning process not only ensured effective plant removal but also left the dead plants in the row, where they could decompose and enrich the soil as natural humus. Additionally, the electric thinning machine minimized the need for manual labor and the use of chemicals, offering an environmentally friendly alternative to traditional crop-thinning methods.

In summary, the Electric Row-Crop Thinning Machine automated and enhanced the process of row crop thinning, allowing for efficient, high-volume, and selective plant blocking. This approach enabled healthier crop growth while reducing labor, time, and environmental impact.

The "Electric Weed and Insect Control in Crop Rows" apparatus

Earl Cecil Rainey's invention for "means for electrically destroying undesired plant life along crop rows" provided a novel solution by using electric currents to eliminate unwanted plant and insect life (Fig. 2.5). This machine targeted weeds and pests that emerge around crops, effectively managing these threats at critical growth stages (Rainey, 1954).

One of the main objectives of this invention was to thin and protect plants by applying precise, high-voltage electric charges to weeds, grasses, and insects. This technique enabled farmers to selectively destroy unwanted vegetation and pests while leaving healthy crop "hills" at consistent intervals. These hills, clusters of one or more plants spaced out along the row, were left untouched by the electric currents, promoting even spacing for crop growth.

Typically, grass and weeds grow faster than the crops and can choke out young plants. Additionally, planting often disturbs the soil, bringing insect eggs to the surface, where they hatch and thrive within the protective cover of the newly sprouted weeds and grasses. Rainey's machine used electric currents to eliminate both the weeds and the pests hiding among them, attacking these threats during early growth stages. Later, when weeds inevitably reappeared between the crop hills, the machine could be employed again to clear them out without harming the established plants.

The machine was designed to be driven along the crop rows, applying high-voltage current selectively. A mobile carrier supported an electric power source and a discharge mechanism, which directed current through plants to the ground. This discharge mechanism included a vertically mounted post with a "star wheel" attached at its lower end. This star wheel had radial prongs that extended across the row, spaced to engage plants at regular intervals. These prongs were grooved on the underside, creating housings for small discharge elements that were electrically connected to the power source. As the machine moved along the row, the prongs made intermittent contact with the weeds and pests, delivering lethal electric pulses that destroyed them without damaging the crop hills.

The spacing of the electric pulses was carefully managed, with wiper rings and wiper fingers mounted to the star wheel, ensuring the current discharged only when the prongs touched undesired plants or insects. This method conserved energy while targeting the areas most in need of treatment. As a result, crop rows

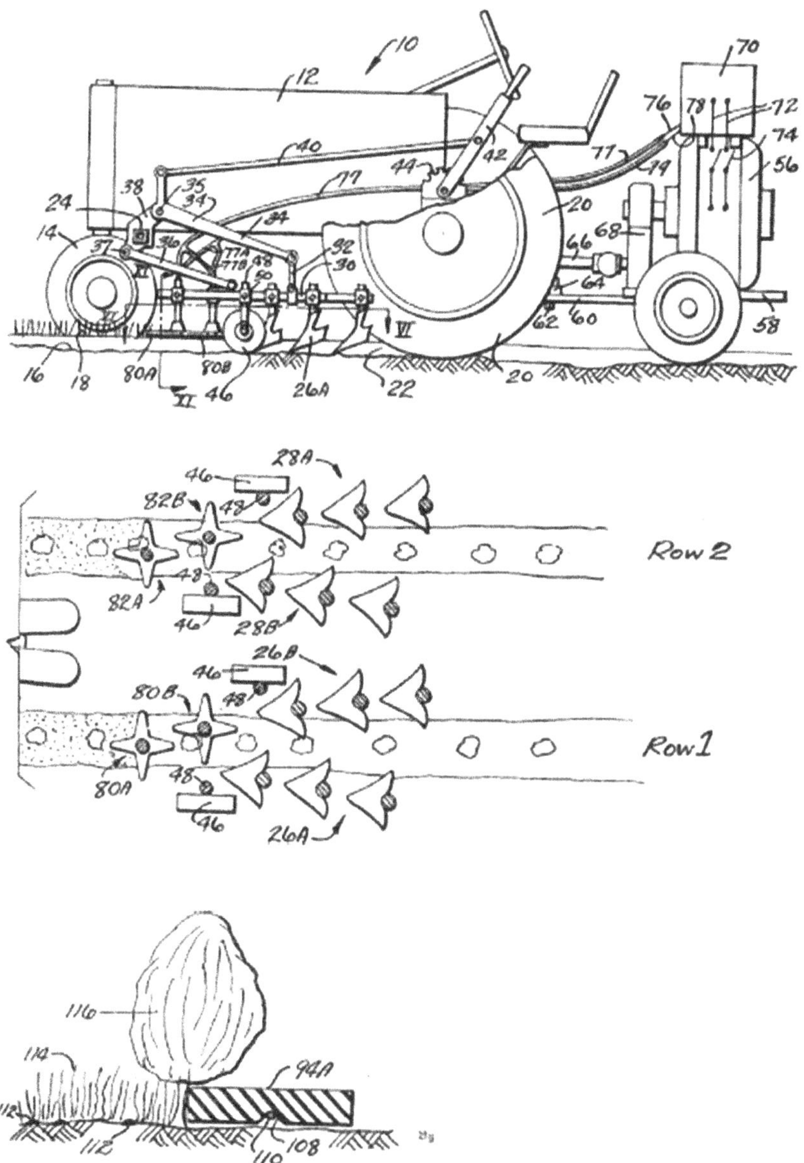

Fig. 2.5. The "Electric Weed and Insect Control in Crop Rows" apparatus. (Rainey, 1954).

could be cleared more efficiently than with manual labor or conventional mechanical thinning, reducing both labor costs and the environmental impact of chemical herbicides.

This EWC machine offered numerous advantages for row crop cultivation. It not only reduced the need for labor-intensive hand thinning but also provided a more uniform spacing of plants, allowing them to grow without interference. Additionally, by destroying insects hiding among the weeds, the machine helped control pest populations that might otherwise have infested the crops. The innovation here lies in its ability to automatically apply targeted electric charges along a row, effectively controlling weeds and insects while enhancing crop yield potential.

By combining selective electrical discharge with automated movement along crop rows, Rainey's invention simplified and optimized the process of maintaining healthy, well-spaced plants. This approach not only protected crops but also contributed to more sustainable farming by reducing dependency on herbicides and manual labor.

Lasco: The first height selectivity electrical weeding apparatus

In 1974, Lasco's Ricks H. Pluenneke proposed a method and apparatus designed for electrically eradicating weeds that operated by leveraging the height difference between crops and weeds within crop rows (Fig. 2.6). The apparatus included adjustable mechanisms specifically tailored to make direct contact with weeds while sparing the crops. These mechanisms could consist of deflectable spring-like elements that were carefully calibrated so that those passing over crop rows were set above the tallest crop plants, while those operating around the crop rows are positioned just above the ground. This configuration minimized the risk of arcing between the apparatus and the ground, ensuring efficiency in targeting weeds without unintended electrical discharge (Dykes, 1978).

The machine itself was composed of a mobile vehicle equipped with a high-voltage

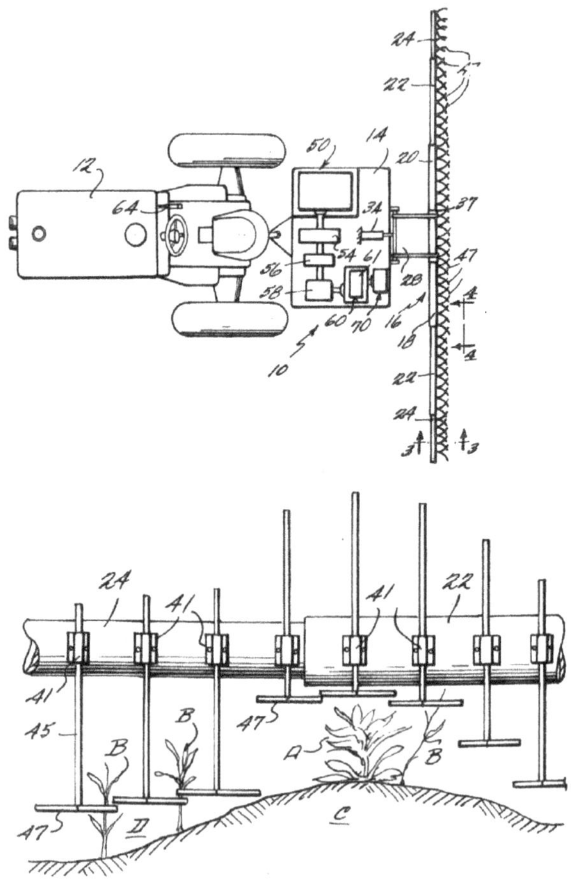

Fig. 2.6. Lasco: The first proposed height selectivity electrical weeding apparatus.

electricity source that can be grounded effectively. It featured an electrically conductive rod mounted in a generally horizontal position and connected to the high-voltage source, designed to be parallel to the ground. This rod, responsible for transferring electricity directly to the weeds, was outfitted with a series of conductive members along its length, each consisting of paired spring-like leaves. These elements were spaced and arranged to form a continuous barrier capable of destroying any plants in their path. Each member could be individually adjusted in height relative to the ground, keeping them high enough to avoid grounding while ensuring maximum contact with target weeds.

For added flexibility, the height of each conductive member could be modified with a vertical bar, enabling the user to fine-tune the apparatus according to variations in terrain or crop height. This adaptability extended to the rod itself, which could be adjusted transversely to reach areas further from the main vehicle path. To further enhance reach, the rod was constructed with telescopic segments that allowed it to extend and retract as needed. This modular design enabled the apparatus to accommodate different crop row widths and plant heights.

A series of safeguards ensured the apparatus operated only under safe conditions. Grounding is achieved with a conductive, sharp-edged wheel that penetrated the soil, connecting both the vehicle and high-voltage source to the ground. Safety mechanisms prevented electricity from being transferred to the rod if the vehicle is not traveling at or above a certain speed, maintaining safe operation during transit. Additionally, if the ground connection was insufficient, the apparatus would halt the delivery of electricity, thereby preventing accidental discharge. Other safety protocols included mechanisms that prevented the apparatus from operating if the access cover to the high-voltage source was open, or if the operator-controlled switch had not been engaged.

In cases where taller plants needed removal, the apparatus could deploy an electrode system to deliver electricity below the soil surface, directly impacting perennial weed root systems. This electrode, partially insulated above ground and uninsulated below, was inserted vertically into the ground, targeting deeper weed growth and enhancing overall weed control.

This weed-killing apparatus could be adjusted for various crop row heights, offering a robust and flexible solution for agricultural weed management through carefully controlled high-voltage application.

Diprose

Dr Diprose conducted laboratory and field research to determine the effectiveness of electrocution on annual weed beets infesting the sugar beet crop and found that effective control of sugar beet bolts required in excess of 5 kV in order to avoid excessively long treatment times (Diprose *et al.*, 1980). One field study that contained a mobile generating unit found that the treatments of 4, 6, and 8 kV were effective regardless of contact time, whereas 3 kV was effective only at contact times greater than 5 s.

Zasso's (Sayyou) foundational work in 1988: "The Use of Electrical Discharge to Control Weeds"

In 1988, Fernando Marques de Almeida presented a thesis to Universidade Estadual Paulista (UNESP), in Botucatu, Brazil. In this thesis, he presented the first complete work on the matter. He studied the energy needed to kill different plant species, how the parameters of the application would influence the results of an application, and the potential economic viability of the system. For the first time, there was a real possibility of having economically viable electrical weeding equipment on the horizon.

Moreover, in this foundational work could be seen the first calculations of future viability of electrical weeding when compared against chemical weed control, as per Table 2.1, extracted from the 1988 work (Almeida, 1988).

The "Weed Electrifier"

In 1997, Louis Charles Strieber proposed the "Weed Electrifier" (Fig. 2.7; Strieber, 1998). The invention presented a unique hand tool specifically designed for trimming vegetation while

Table 2.1. "The Use of Electrical Discharge to Control Weeds" cost comparison between electric and chemical weeding. (Almeida, 1988).

ItemBase: cruzados from February 1988	Chemical control (Cz$)	Electrical discharge (Cz$)
Depreciation		
Tractor	114.46	385.98
Electrocution equipment	—	480.54
Sprayer	38.60	—
Fuel		
Tractor	53.10	179.05
Electrocution equipment	—	767.34
Lubricants	10.62	189.05
Repair and maintenance	17.01	96.29
Housing, insurance, taxes	3.40	19.26
Herbicide	5,133.60	—
Labor	67.80	271.44
Remuneration		
Fixed capital	49.56	286.36
Working capital	489.10	135.31
Total	5,976.53	2,810.62

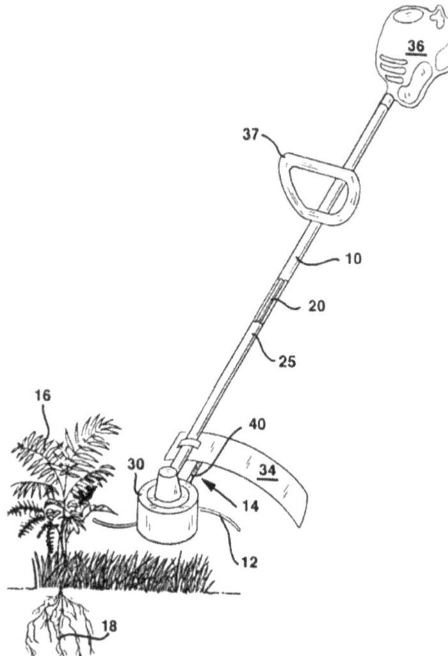

Fig. 2.7. The "Weed Electrifier." (Strieber, 1998).

simultaneously applying an electrical current. This tool integrated a generator positioned within or near the trimmer head, which harnessed the rotational energy of the drive shaft—used to spin the cutting blade—to produce electricity. This electricity flowed through the trimmer, electrifying the vegetation as it is cut, adding an additional dimension of effectiveness to the trimming process.

Additionally, the invention included a novel add-on kit designed to transform a standard weed trimmer into an electric weed-killing device. This kit was easy to install and remove, enabling users to convert their existing equipment into an electrified trimmer without the need for specialized machinery. By applying electrical treatment to weeds, the need for repeated trimming or the use of chemical poisons was reduced, saving both time and resources while minimizing environmental impact. Even low-powered trimmers could generate sufficient electricity with this kit, as most gasoline-powered models provided more than enough power to drive both the cutting mechanism and the attached generator.

One of the significant benefits of this tool was its ability to clear dense brush, including larger invasive plants and trees, without relying on chemical treatments. As the tool electrified the weeds, electricity flowed through the plant's fluids down to the roots, killing the vegetation more thoroughly. Because weeds generally grow taller than cultivated plants, the tool was able to target them selectively, sparing other plants with minimal physical contact. For more robust plants, such as woody weeds, mesquite, and prickly ash, the tool could be applied directly at the base of the plant, cutting through the bark and reaching the cambium layer to prevent regrowth from the trunk.

Optimal conditions for this treatment occurred after a rainfall, as the moisture enhanced the conductivity of the soil, allowing the electrical current to flow down through the roots and further into the surrounding ground, effectively disrupting the weed's regenerative capabilities. This process was most effective when applied multiple times, as seeds that may have dispersed before the initial treatment would sprout into weeds that could then be targeted in subsequent rounds of electrification. Larger or more resilient weeds may have required additional treatments to fully prevent regrowth, making this method a preferable alternative to repeated physical trimming or the application of chemical herbicides.

Safety was paramount when operating this electrified tool. In addition to standard eye and ear protection, the operator should wear rubber gloves, boots, and protective chaps to mitigate any risk of electrical shock. A conscientious operator would also conduct regular inspections of the tool to ensure its components, such as the cutting head, brushes, teeth, and hub, were in optimal condition, as wear and imbalance could compromise both safety and effectiveness.

The advantages and innovative aspects of this invention were apparent in its ability to manage vegetation without chemicals, making it a versatile, environmentally friendly solution for weed and brush control.

The "Zero Weed"

In 2001, Bertil Persson, Par Henriksson, Tomas Nubrant, and Beritt Mattsson proposed the "Zero Weed" (Fig. 2.8; Persson *et al.*, 2001). Despite known challenges, the invention aimed to introduce a method for weed control that relied on the application of electricity. Remarkably, effective weed management could be achieved with minimal energy consumption by using brief, high-voltage pulses. These pulses, though low in energy content, were sufficient to penetrate the cell membranes of young, sprouting weeds. This approach to weed control proved beneficial across a range of applications, including agriculture, horticulture, forestry, and park maintenance, particularly when used in conjunction with sowing. Notably, this electrical treatment impacted only actively growing seeds and early-stage plants, inhibiting their development without affecting dry seeds intended for future germination. This feature made the invention ideally suited for simultaneous application during sowing, as it did not interfere with the intended crop's seeds or disrupt the soil's microflora within the specified voltage range.

Young seeds, in the early stages of growth, display heightened sensitivity to electrical currents. However, once a seed develops into a plant, significantly more electrical energy is required to halt its growth permanently. For ungerminated seeds, substantial energy is needed to the extent that it would essentially "boil" the seed. The problem lies in the fact that by the time a desired crop is sown, weed seeds already present in the soil may have begun to germinate. Traditional techniques necessitate weed control to allow the intended crop to flourish unimpeded.

By applying low-energy electrical pulses to soil prepared for sowing, the germinating weeds could be stopped in their development. When this treatment coincided with sowing, the crop gained a valuable head start, sprouting before weeds had the opportunity to establish themselves. This simultaneous approach provided the crop with a competitive advantage, reducing the chance of later-emerging weeds overtaking the field. Additionally, because the ground had often undergone prior mechanical preparation, any preexisting weeds were likely already removed. Short, high-voltage pulses directed into the soil could effectively control these newly sprouting weeds. While these pulses were brief, the energy level was sufficient to permeate the cell membranes of young weed sprouts, thereby arresting their growth.

The crop seeds, still ungerminated at the time of sowing, remained unaffected by this electrical exposure. This dual action allowed weed control and sowing to occur with the same equipment, presenting a practical and time-efficient solution. By delivering weed suppression at the

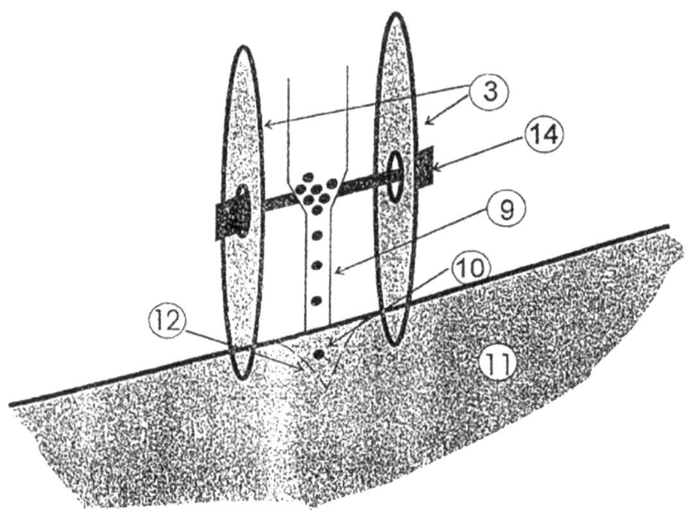

Fig. 2.8. The "Zero Weed." (Persson *et al.*, 2001).

optimal moment, competition from weeds was minimized precisely when the crop was beginning to establish itself. This method of weed control required minimal energy input, as demonstrated by a straightforward power formula that calculated mean power as a function of soil conductivity, field strength, modulation degree, and material density.

Although the pulse's momentary power level was high, typically around 12 MW/kg, the brief duration of each pulse—often only 100 μs—meant that the overall mean power was significantly lower, at about 12 kW/kg. In practice, a charged capacitor battery could discharge quickly to provide these high-voltage pulses. An alternative approach involved using alternating voltage transformed to a high level.

The impact of these high-voltage pulses on plant cells primarily affected the protoplasm, with the cell wall acting as a critical barrier. When rectangular pulses were applied, optimal field strength ranged between 100,000–300,000 V/m, with pulse durations of 10–100 μs. With exponentially decaying pulses, the field strength may have been lower, between 25,000–75,000 V/m, but the pulse duration extended from 1,000–20,000 μs. This approach offered an innovative and energy-efficient solution for targeted weed management, harnessing electricity to protect crop yields without harming the soil's natural biological processes.

The birth of Sayyou

In 1999, Japanese brothers Satoru and Yutaka Narita, together with Sayyou engineers, proposed many ideas of how to perform electrical weeding in different scenarios. Moreover, they also studied the effect of AC frequency and DC current in the performance, efficiency, and efficacy of electrical weeding. This was the foundational patent of their company, Sayyou (Figs. 2.9 and 2.10), which led the development of the field until it changed control and name to Zasso, who ultimately created the first economically viable electrical weeding equipment.

The electrode commutation idea

In 2006, Sayyou's Constantino Augusto Henrique Schwager and his son, Jeancarlo Ricardo Schwager proposed the electrode commutation idea for an electrical weeding device (Fig. 2.11; Schwager and Schwager, 2005).

The invention presented an electromechanical device specifically designed for the eradication of noxious weeds using high-voltage discharges delivered through an electrode system subdivided into multiple smaller electrodes, known as multiple electrodes. Each of these electrodes operated within a voltage range of 1,000–35,000 V,

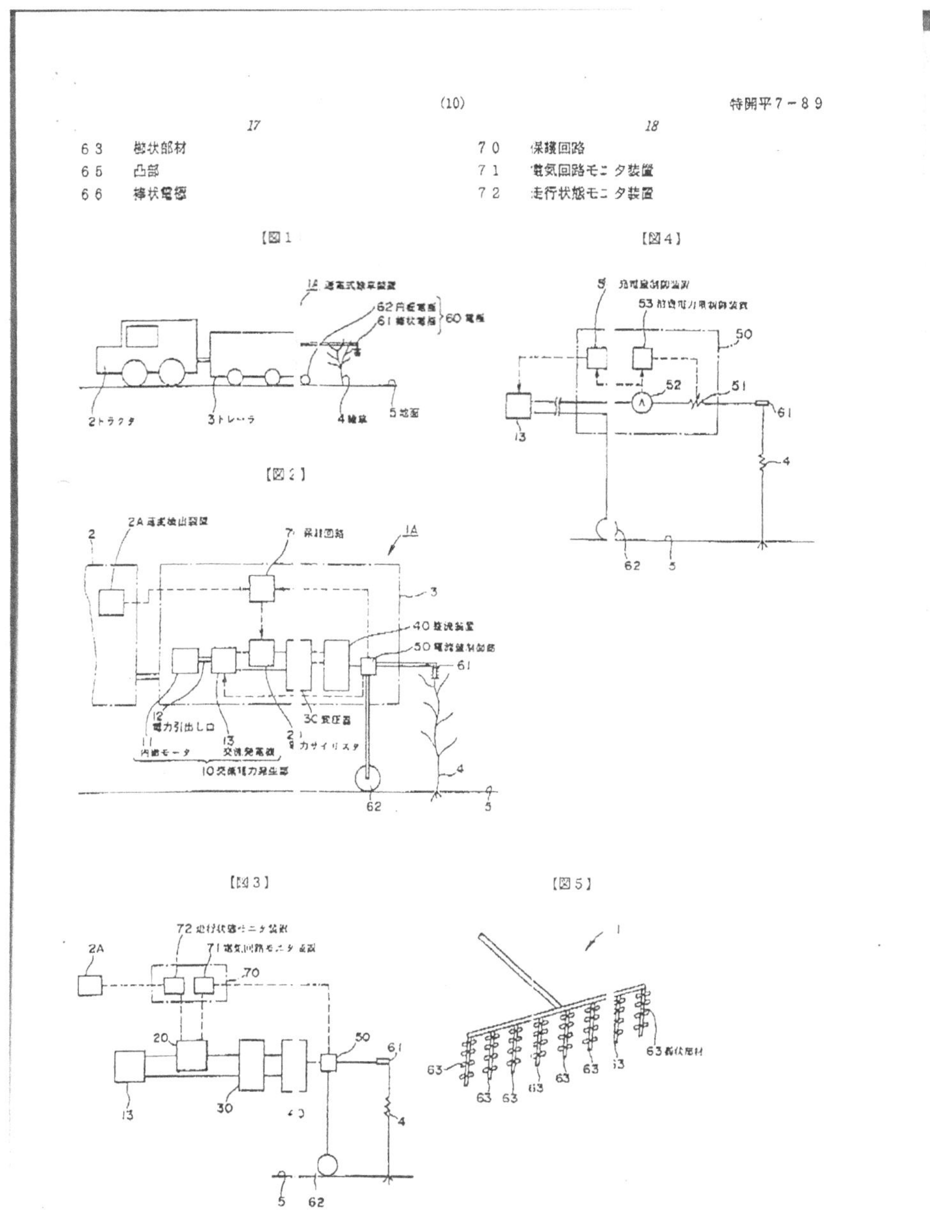

Fig. 2.9. Design excerpt from 1999's first Sayyou patent.

controlled by an electronic commutation system that alternated between conduction and cutoff periods of electric energy. These intervals, lasting between 0.–2.0 s, allowed each electrode to operate efficiently without continuously drawing power.

The system's design strategically reduced the number of plants receiving the electric discharge

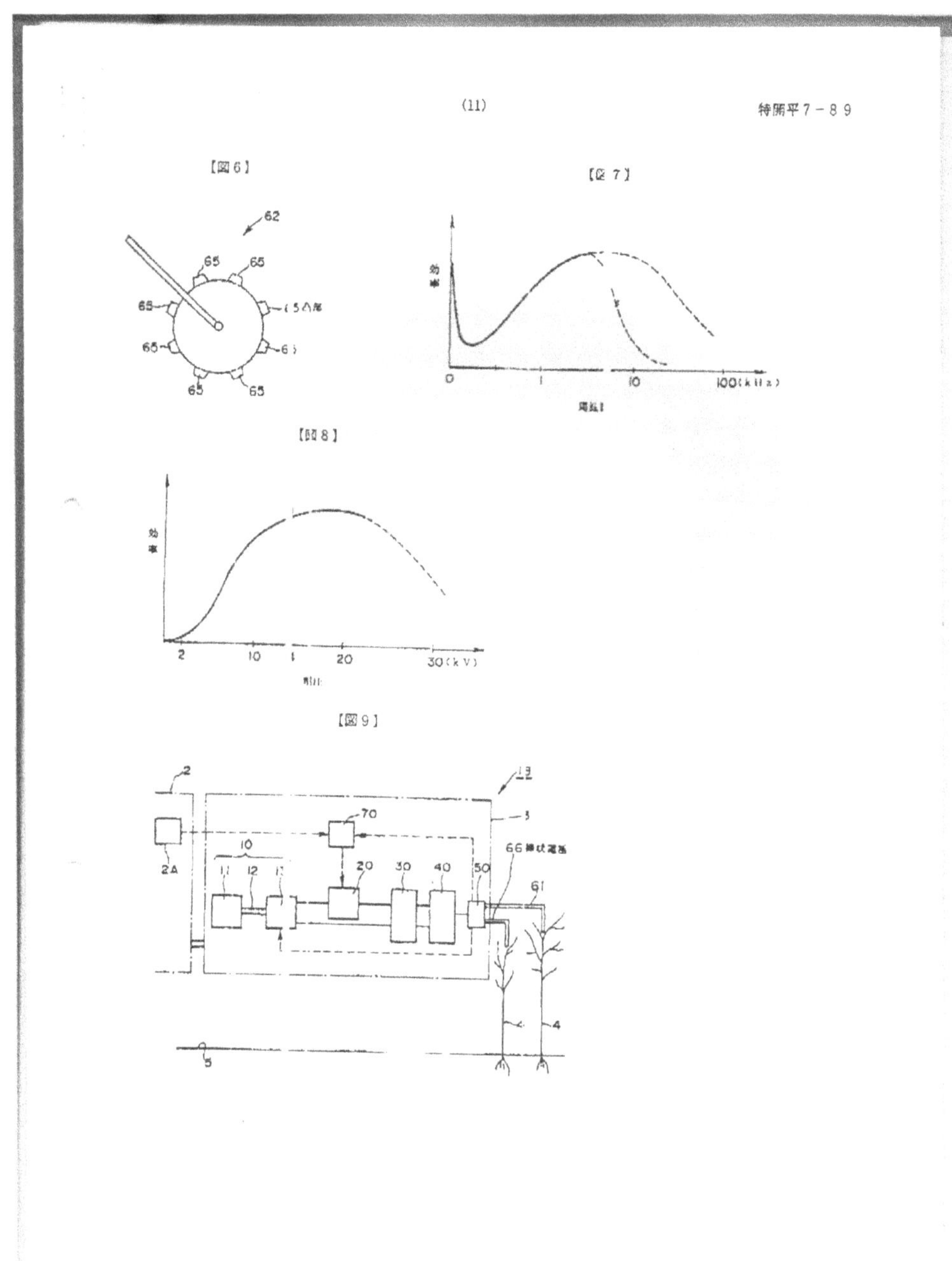

Fig. 2.10. Design excerpt from 1999's first Sayyou patent.

at any given moment. By distributing the discharge across separate electrodes and managing the timing, the device reduced the overall power demand on the generator, improving the efficiency and output of the weed control process. The device's power, whether in DC or AC, was generated by a low-voltage generator and elevated to the required high voltage through a transformer, followed by a current rectifier. From there, an electric-electronic distributor channeled the energy to each of the multiple electrodes, ensuring a steady supply and controlled delivery of high-voltage pulses.

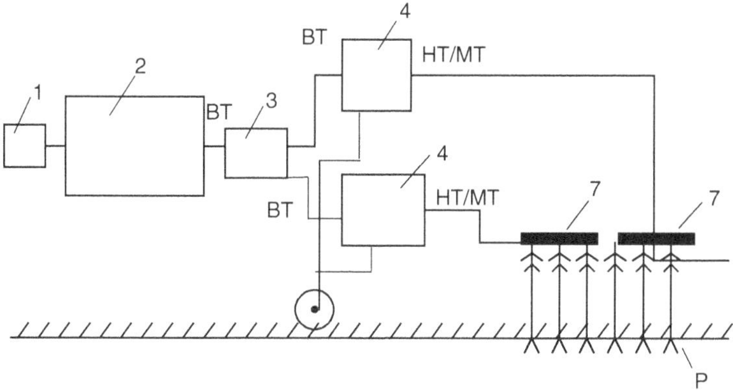

Fig. 2.11. The electrode commutation idea.

An essential feature of this design was its ability to prevent the formation of electric arcs, which could otherwise lead to fires. By precisely controlling the periods of conduction and cut-off in the energy flow to the electrodes, the device ensured that energy was applied in a measured way, reducing the risk of unintended arcing. This careful regulation not only enhanced the safety and reliability of the device but also optimized its performance, making it a powerful tool for selective, high-voltage weed eradication.

The high frequency idea

In January 2013, in Sayyou's filed and discontinued patents of BR 13 2013 011223 1, BR 13 2013 011222 3, BR 13 2013 011208 8, and BR 10 2013 007771 2, different equipment embodiments were proposed that used, for the first time, high frequency to control weeds (Fig. 2.12). The embodiments included both manual and tractor-based equipment. The first of these patents used the knowledge from Japanese Patent No. 19991130 of 1992, where it is thought that the efficiency of the application varies according to frequency, reaching higher efficiency at values above 1 kHz or at DC (Coutinho Filho, 2013).

BR 13 2013 011223 1 reads:

"... *the 'HIGH FREQUENCY EQUIPMENT FOR ELECTROCUTION OF PLANTS IN LARGE VOLUME' was developed, which is understood by an "a" Chassis, which is the support structure of the equipment that unites all the other components in monolithic form.*

There is also the Alternator, "b", [which] comprises any type of alternator, generator or mechanical energy converter that converts it into electrical energy. The device in question can be powered (by a tractor or similar) or by an integrated motor. In this case, it may be necessary for the equipment to be automotive.

Regarding the controller device "c" of the equipment, it comprises an electronic-electric circuit that: (1) activates the system; (2) controls the input current of the frequency converter; (3) controls the input voltage of the drive; (4) controls the frequency of the drive.

Regarding the frequency converter, "d" is an integrated circuit (Flyback or similar) that: (5) increases the frequency of the system; (6) reduces the cost of production, since the cost of the transformer may be the highest single cost in production and is directly linked to the amount of material expended for it; (7) reduces the weight of the transformer as a consequence of the above relationship, making transport easier and less energy is spent in the displacement, thereby increasing the energy efficiency of the system.

This increase in frequency allows the use of voltage transformers of much lower weight than the systems used for this purpose to this day. The maximum power supported by a transform[er] is directly proportional to the frequency, so the higher the frequency, the greater the maximum power in the same transformer (of course, this fact is restricted to the

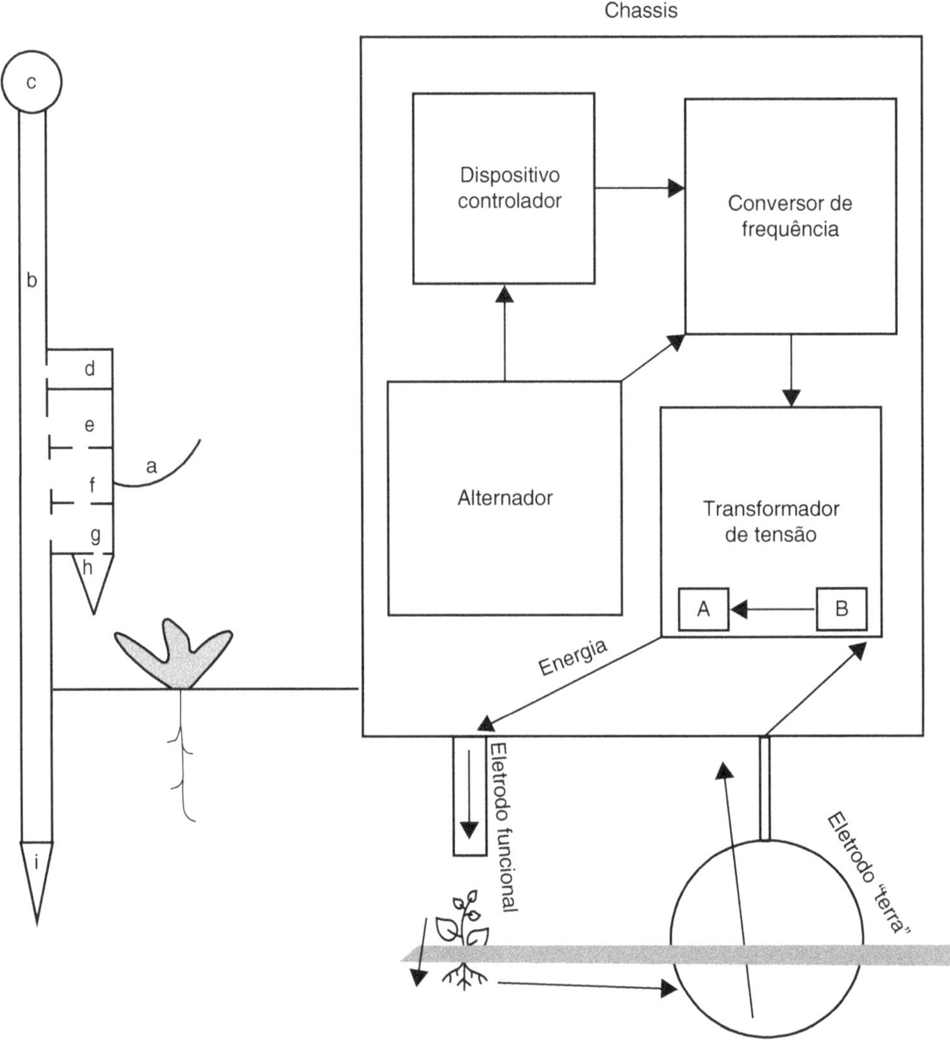

Fig. 2.12. Design excerpts from the first 2013 patents BR 13 2013 011223 1, BR 13 2013 011222 3, BR 13 2013 011208 8, and BR 10 2013 007771 2 showing the concept of inverters being used to increase alternating current (AC) frequency for electrical weeding.

calculation of the transformer, its core, the number of turns, its characteristics Physical and etc...).

The output frequency is greater than or equal to 500 Hz, as shown in Japanese Patent No. 19991130 of 1992, where the efficiency of the application varies according to frequency, reaching higher efficiency values from 1 kHz or with direct current.

The frequency converter can be modular or unique for all equipment. Being modular allows for easier adaptation to power requirements, and being unified can mean lower production costs.

Such a converter is the main factor that differentiates this equipment from the other deposited and similar ones used and developed in the past, being its main innovation.

With respect to the voltage transformer "e", the estimated average voltage is approximately 6,800 volts and, except in specific cases, it may not be less than 220 Volts, or greater than 100 Kilovolts."

The impedance matching idea

In 2018, Zasso proposed the idea of impedance matching in the patent BR 10 2019 002353-8 (Rona *et al.*, 2019). This invention presented a groundbreaking weed inactivation device that functioned as a physical herbicide, utilizing high-voltage electrical currents to incapacitate weeds without the environmental risks associated with chemical herbicides. By applying electric discharges directly to the target plants via electrodes, the device avoided the use of potentially harmful chemicals that could seep into the soil and food chain. The concept of physical herbicide here relied on high voltage, ranging between 1–20 kV, delivered directly to or near the weeds, effectively neutralizing them in a more sustainable manner.

At the core of this technology was an advanced electronic topology—a high-power-factor converter built from simple and widely available electronic components. This converter managed the power supply without the need for complex software or bulky components common in older technologies. The primary components of the converter included an inverter, an inductive or capacitive harmonic filter, and a capacitive voltage multiplier composed of diodes and capacitors. These elements worked in tandem to convert and control power efficiently, supporting the device's weed inactivation function seamlessly.

The power conversion process began when a DC or AC power source generated an initial voltage, which was then passed through an inverter that increased the frequency of the current. This high-frequency current then moved through the harmonic filter, which maintained a high-power factor. By diminishing or even eliminating the need for a separate power factor correction unit, the harmonic filter made the overall device more compact and energy efficient. In some configurations, a high-frequency transformer followed the harmonic filter, further amplifying the voltage for the next stages of the device.

The voltage multiplier, which could be of the Cockcroft-Walton type or a similar design, then took this high-frequency, high-voltage output and multiplied it further, achieving the higher voltage levels needed for effective weed inactivation. This multiplier adapted to varying loads automatically, allowing for dynamic power adjustment without any additional control systems. Such adaptability ensured that the device could maintain effective output across different conditions and weed densities, making it both powerful and versatile for agricultural applications.

An emerging benefit of this device was its inherent impedance-matching capability, which was a by-product of the voltage multiplier. By providing a series impedance at the transformer's secondary, the voltage multiplier prevented the transformer from operating in an open-circuit state, reducing resonance risks that could otherwise have caused damaging voltage peaks or insulation issues. This design also allowed the inverter to operate in a resonant or quasi-resonant mode, which minimized harmonic interference, preserved transformer integrity, and improved the overall efficiency of the system by lowering conduction losses in the inverter's switching components. This impedance-matching feature was integral to the system's performance, enabling a safer, more reliable, and highly efficient means of weed inactivation.

This meant a translation from high-frequency AC back to DC output. The DC may have had AC components in the form of ripples, both in the range of the alternator (usually 3x the frequency of the alternator, if it is triphasic), and possibly also parasitic high-frequency components, due to harmonics.

A simple embodiment is shown in Fig. 2.13.

Introduction

As electrical weeding evolves into a scalable and energy-efficient alternative to chemical herbicides, the challenge of delivering stable, high-voltage power across variable soil and plant conditions becomes paramount. The input parallel output series (IPOS) architecture presents a groundbreaking solution that optimizes both power delivery and modularity in electrical weeding devices (Zasso Group AG and Serviço Nacional de Aprendizagem Industrial – SENAI, 2024).

This section presents the principles, construction, advantages, and control methodology of IPOS-based converters applied to EWC, as disclosed in Patent Cooperation Treaty (PCT) patent application BR2024050379.

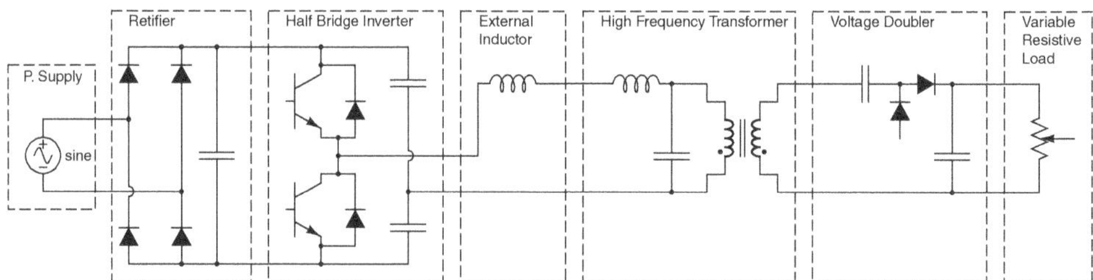

Fig. 2.13. Design excerpts from the first 2018 patents, EP2018/062271. The input parallel output series idea

The problem: Dynamic load variability in electrical weeding

Electrical weeding efficacy depends on delivering lethal energy to a plant's root system, which varies significantly due to weed density and type (weed pressure), soil conductivity (moisture, compaction, mineral content), and electrode–soil contact quality.

These variables produce constantly changing load resistance, ranging from 0.5 kΩ to 35 kΩ, across which power must be consistently delivered. Traditional solutions struggle with this dynamic variability, leading to uneven application, wasted energy, overdimensioned components, voltage peaks that reduce safety and efficiency, and other such issues.

A solution: Input parallel output series

The IPOS architecture is a power conversion strategy specifically designed to address the challenge of maintaining high-performance operation in systems that require both high voltage and adaptability to dynamically changing loads, as is the case in electrical weeding (Fig. 2.14). The approach begins with its input parallel configuration, in which multiple power cells receive current from a shared DC bus. This shared arrangement evenly distributes the current draw across the entire system, thereby reducing stress on individual components. It also helps ensure that no single power cell is overburdened during operation, which enhances reliability, extends component lifespan, and simplifies thermal management due to lower localized heating.

At the same time, the output series structure is used to aggregate the voltages produced by each individual power cell. Because each power cell contributes a fixed amount of DC voltage, placing them in series results in a cumulative effect, allowing the system to achieve a much higher overall output voltage. For example, five power cells each providing a regulated 2.5 kV can be stacked to generate a combined output of 12.5 kV. This modular voltage buildup is ideal for agricultural applications where the electrical load varies significantly depending on the type of weeds, the density of the vegetation, and the resistance of the surrounding soil.

Key benefits of this topology include high-voltage gain without the need for high-voltage handling in each individual module, making it easier to maintain electrical insulation and component safety. The system is inherently scalable; additional power cells can be added to increase voltage output, or removed to reduce it, without altering the control logic or physical design significantly. Furthermore, the IPOS layout offers built-in redundancy. If one module fails, others can continue operating, and maintenance can be performed on a per-cell basis. This modularity dramatically simplifies field service and troubleshooting in complex or remote agricultural environments.

The power conversion architecture in the IPOS system is built around the use of modular power cells, each equipped with a complete set of subsystems necessary for isolated power transformation and voltage regulation. Each power cell begins with an inverter switching arm, which typically operates as a half-bridge within a phase-shift full bridge (PSFB) topology. The PSFB configuration allows precise regulation of the power conversion process while enabling soft switching, which reduces switching losses and electromagnetic interference.

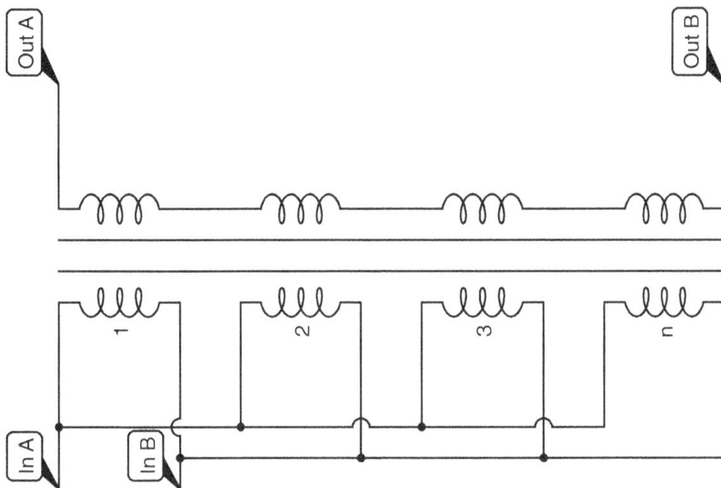

Fig. 2.14. Input parallel output series (IPOS) example where every cell is composed of only one transformer.

The inverter outputs a high-frequency AC signal that feeds into a high-voltage transformer. This transformer serves a dual purpose: it isolates the input from the output to improve safety, and it steps up the voltage to the required range for effective plant electrocution. The output of the transformer is then passed through a full-wave rectifier, which converts the AC signal back into DC.

After rectification, the signal is cleaned using an inductive-capacitive filter, which smooths out any residual voltage ripple and suppresses noise. This filter is essential for maintaining a consistent DC output that is safe for both plants and electronics. Each cell also integrates gate drivers, crossover protection circuits, and sensors for current and voltage monitoring, ensuring reliable operation and safe handling of faults.

What unites the system is the use of a shared switching arm, which forms the second half of the full-bridge topology for all power cells. This shared component, managed by the central control unit, synchronizes all switching actions and phase shifts to regulate the total power output with high precision.

In field conditions, where plant and soil resistance vary constantly, maintaining consistent power delivery is crucial for ensuring effective weed electrocution. The IPOS system solves this by integrating a dynamic feedback loop into the control architecture. It begins by assessing the weed pressure, which may be estimated either manually or via computer vision systems that analyze vegetation density and type in real time.

Based on this analysis, the control unit determines the target power level that needs to be delivered. Rather than holding voltage constant and letting power vary with resistance—an approach that leads to uneven performance—the system adjusts the phase shift between the inverter switching arms and the shared arm, thereby regulating voltage output in real time. The pulse width modulation (PWM) duty cycle of the gate drivers is also dynamically tuned to fine-tune power delivery.

The result is constant power delivery across a wide range of resistances. For example, as shown in Table 2.1, a 6 kW output can be maintained whether the system faces a 3 kΩ load (requiring 4242.6 V and 1.414 A) or a 15 kΩ load (requiring 9486.8 V and 0.632 A). If resistance increases beyond safe operating conditions, the system intelligently derates itself, reducing power output (e.g. to 80%) to protect components and maintain system integrity.

Controlling an input parallel output series system

At the heart of the IPOS system may (there are options without such controls) lie a centralized control unit responsible for orchestrating power delivery. This unit performs several interdependent tasks. First, it manages the phase-shift timing across all power cells, synchronizing the switching of each cell's inverter arm with the shared switching arm. This timing determines the effective voltage contributed by each cell and, by extension, the total output. Second, the control unit interfaces with gate drivers, sending activation signals in the form of PWM waveforms. These signals regulate the timing and duration of each switching cycle, allowing the system to respond instantaneously to changes in load resistance or weed pressure.

To monitor output conditions, the control unit receives continuous input from Hall effect current sensors and resistive voltage dividers installed in each power cell. These data are used to verify that each cell contributes the appropriate voltage and that the system remains within safe operating limits. The use of equipotential balancing through voltage feedback ensures that no one power cell bears an undue burden, which could lead to thermal stress or premature failure.

Beyond performance control, the feedback loop also incorporates safety functions, including spark detection and derating logic. These features ensure that the system disables or limits output when voltage arcs occur or when environmental conditions change abruptly, preventing fire risks and system damage.

The advantages of an input parallel output series system in electrical weeding

The IPOS system delivers constant power output across varying electrical loads, something essential in environments with dynamic soil and plant resistance. It responds in real time to changes in load impedance by adjusting voltage rather than current, preserving both energy efficiency and application consistency.

It achieves high efficiency through phase-shifted switching that allows zero voltage switching (ZVS) and zero current switching (ZCS). These methods reduce switching losses and extend the life of switching components, contributing to an overall reduction in operational cost and complexity.

As a modular and scalable system, IPOS allows voltage to be scaled simply by adding or removing power cells, with minimal reconfiguration. This modularity also makes the system easier to maintain, since faulty cells can be replaced individually without dismantling the entire setup.

The system's compact and cost-effective design minimizes insulation requirements, as each cell operates at a lower internal voltage than the full output voltage. This makes IPOS suitable for a wide range of applications, from handheld devices to high-power vehicle-mounted systems.

The intelligent control system, with feedback from sensors and vision-based weed detection, allows application-specific power tuning. This prevents overapplication of power in low-density weed zones, reducing energy waste and environmental risk.

Comparison to prior solutions

Compared to prior solutions, IPOS represents a significant leap forward. AC systems using PWM modulation regulate average power only, which allows wide power fluctuation during each waveform cycle. Cockcroft-Walton circuits provide fixed-voltage gain and suffer from instability, parasitic ripples, and lack of load adaptability. Voltage multipliers with passive impedance matching offer some improvement but still lack dynamic control and scalability.

In contrast, IPOS enables constant, digitally controlled power delivery with scalable modularity, achieving low parasitic ripple, high efficiency, and superior adaptability. It is the only system in this context that marries phase-shifted digital control, modular output stacking, and load-responsive feedback into a unified platform suitable for field-scale electrical weeding.

The Continued Technological Journey of Electrical Weeding

The technological journey from early EWC concepts to Zasso's modern innovations illustrates a fascinating evolution driven by advancements in engineering, materials science, and an increasing understanding of plant biology. The quest for an effective, nonchemical weed management solution dates back over a century, with one of the first documented inventions being Albert A. Sharp's "Vegetation Exterminator" in 1893. Sharp's early model, which relied on simple high-voltage application to damage weeds, marked a visionary beginning in EWC. However, despite its promise, limitations in voltage control, power efficiency, and practical field applicability hindered widespread adoption. Sharp's invention laid the foundational understanding of using electricity to disrupt plant growth, sparking curiosity and further experimentation in the decades to follow.

In the early and mid-20th century, advances in electrical engineering began to influence agricultural applications, yet attempts to refine Sharp's original vision met with considerable technical challenges. At this time, controlling and sustaining high voltage without risking electrical arcs or excessive energy consumption proved difficult, particularly in open fields. Moreover, early systems lacked the precision needed to differentiate weeds from crops, limiting their use to situations where the proximity of crops wasn't a concern. Throughout these years, researchers and inventors continued experimenting with various methods to improve power efficiency and control, but the technology remained rudimentary, often requiring extensive manual oversight and remaining expensive to operate on a large scale.

As electronics evolved in the late 20th century, the potential for EWC started to expand. The emergence of solid-state electronics, frequency converters, and more compact power sources provided a fresh impetus for electrical weed management technologies. Innovations in electronic circuits allowed for better control of high-voltage currents, creating the possibility of adjusting power based on weed density and proximity to crops. The integration of transformers, frequency converters, and early impedance-matching systems further advanced this field, allowing for safer and more controlled application of electrical pulses. These systems achieved varying levels of success, but they were often too large, heavy, or inefficient to be practical for routine agricultural use.

Zasso's recent contributions mark a significant leap forward in this lineage of technological development. Building on more than a century of incremental progress, Zasso has introduced high-frequency modular systems that overcome the traditional limitations of weight, power inefficiency, and safety risks. Unlike earlier models, Zasso's technology is specifically designed for versatility in modern agriculture, with components that can be customized to suit different field sizes, weed types, and crop arrangements. Central to Zasso's success is its implementation of advanced frequency converters and voltage multipliers, which enable the equipment to deliver high-voltage pulses precisely and efficiently. By using short, high-frequency pulses rather than sustained current, these systems can achieve effective weed control with minimal energy consumption.

The capacitive voltage multiplier, as employed by Zasso, represents a landmark in the evolution of electrical weed management. This component allows for voltage levels to be dynamically adjusted based on load, which is particularly advantageous in agricultural settings where weed density and distribution can vary greatly. This self-adjusting feature eliminates the need for complex external controls, a significant improvement over previous generations that required manual recalibration and software-based adjustments. Additionally, the impedance-matching properties of the voltage multiplier protect the equipment from resonance issues that previously plagued high-voltage systems, reducing the risk of electrical peaks that could damage insulation or create dangerous arcing. This innovation not only improves safety but also extends the life and reliability of the equipment, making it a viable, long-term solution for farmers.

Zasso's latest designs also reflect a growing awareness of the environmental and health benefits of nonchemical weed control. As regulatory pressures and consumer preferences increasingly discourage chemical herbicides, Zasso's technology provides a sustainable alternative that is both effective and environmentally friendly. By controlling weeds through electricity, Zasso eliminates the need for harmful chemicals that can leach into soil and water, thus preserving local ecosystems and reducing exposure risks for farmers and consumers alike. This is especially important as modern agriculture seeks to balance productivity with ecological stewardship, aligning with global sustainability goals.

The technological progression from Sharp's "Vegetation Exterminator" to Zasso's advanced systems underscores a broader trend in agricultural technology: the pursuit of efficiency, precision, and environmental responsibility. What began as an ambitious but impractical idea in the late 19th century has been refined through decades of trial, innovation, and scientific breakthroughs. Each generation has built upon the insights and shortcomings of its predecessors, culminating in a technology that is not only functional but also essential for the future of sustainable agriculture.

References

Almeida, F.M. (1988) *O uso de descarga elétrica no controle de plantas daninhas*. Universidade Estadual Paulista (UNESP), Botucatu, Brazil.

Burt, W.E. (1928) *Weed Destroyer. U.S. Patent No. 1,661,030*. U.S. Patent Office, Washington, DC, USA.

Coutinho Filho, S.A. (2013) *Equipamento de Alta Frequência para Eletrocussão de Plantas em Grandes Volumes. Patent No. BR 13 2013 011223 1*. Instituto Nacional da Propriedade Industrial (INPI), Rio de Janeiro, Brazil.

Diprose, M.F., Benson, F.A. and Hackam, R. (1980) *Electrothermal Control of Weed Beet and Bolting Sugarbeet. Weed Research*. Blackwell Scientific Publications, Oxford, UK.

Dykes, W.G. (1978) *Method and Apparatus for Using Electrical Current to Destroy Weeds in and Around Crop Rows. U.S. Patent No. 4,094,095*. U.S. Patent Office, Washington, DC, USA.

McCreight, N.L. and McCreight, J.H. (1953) *Electric Row-Crop Thinning Machine. U.S. Patent No. 2,632,285*. U.S. Patent Office, Washington, DC, USA.

Persson, B., Henriksson, P., Nybrant, T. and Mattsson, B. (2001) *Method and Device for Weed Control. U.S. Patent No. 6,237,278*. U.S. Patent Office, Washington, DC, USA.

Poynor, R.R. (1954) *Electric Weed Killer. U.S. Patent No. 2,682,729*. U.S. Patent Office, Washington, DC, USA.

Rainey, E.C. (1954) *Means for Electrically Destroying Undesired Plant Life Along Crop Rows. U.S. Patent No. 2,687,597*. U.S. Patent Office, Washington, DC, USA.

Rona, S.A., Valverde, B., Souza, D.T.M. de and Coutinho Filho, S.A. (2019) *Dispositivo de Inativação de Plantas Invasoras. Patent No. BR 10 2019 002353-8*. Instituto Nacional da Propriedade Industrial (INPI), Rio de Janeiro, Brazil.

Schwager, A.H. and Schwager, J.R. (2005) *Equipamento de Comutação Eletrônica de Eletrodos Múltiplos para Eletrocussão de Plantas Daninhas. Patent No. PI 0502291-6*. Instituto Nacional da Propriedade Industrial (INPI), Rio de Janeiro, Brazil.

Sharp, A.A. (1893) *Vegetation-Exterminator. U.S. Patent No. 492,635*. U.S. Patent Office, Washington, DC, USA.

Weed Electifier (1998) *Weed Electrifier: Add-on Kit for Weed Eater Converts Trimmer into Electric Weed Eradicator. U.S. Patent No. 5,806,294*. U.S. Patent Office, Washington, DC, USA.

Zasso Group AG and Serviço Nacional de Aprendizagem Industrial – SENAI (2024) *A System and a Method for Powering a Plurality of Electrodes, an Electrical Weeding Device, and, a Vehicle. Patent No. PCT/BR2024/050379*. World Intellectual Property Organization (WIPO), Geneva, Switzerland.

3

Why Plants Die from Electrocution

The Electrical Weeding Process

In electrical weeding, electrodes or electrode arrays are used to deliver controlled electric shocks to the targeted weeds. The electric current is higher where impedance is lower. Plants—as do all living organisms—have water and electrolytes present in their cells, and these substances contribute to their electrical conductivity. Consequently, *ceteris paribus*, a larger portion of the current flowing from the leaves, where the plant contacts the electrodes, reaches the root system, instead of being directly dispersed into the soil. Especially conductive are the xylem and phloem, which work together to transport water, minerals, and nutrients throughout the plant, allowing it to meet its energy and nutritional needs.

Phloem is a type of plant tissue that conducts sugars, amino acids, and other organic compounds throughout the plant. It is made up of sieve tube elements, companion cells, and phloem parenchyma cells. Sieve tube elements are specialized cells that transport sugars and other organic compounds through the plant. They are connected end-to-end by sieve plates, which have small pores through which the materials can pass. Sieve tube elements do not have nuclei, so they rely on companion cells for metabolic support. Companion cells are closely associated with sieve tube elements and provide them with metabolic support. They have their own nuclei and perform functions such as regulating the flow of materials through the sieve tube elements and responding to signals from the plant's hormones. Phloem parenchyma cells are smaller cells that are found interspersed among the sieve tube elements and companion cells. They provide structural support to the phloem tissue and can also store sugars and other organic compounds. Phloem tissue is found in all parts of the plant, including the roots, stems, leaves, and flowers. It plays a crucial role in plant growth and development by transporting nutrients and other materials throughout the plant.

Xylem is a specialized tissue found in plants that is responsible for transporting water and dissolved minerals throughout the plant. It is composed of a series of interconnected tubes and cells that run from the roots to the leaves and is vital for the plant's overall health and growth. There are two types of xylem tissue: primary xylem and secondary xylem. Primary xylem is formed during the plant's early development and is responsible for the upward movement of water and minerals from the roots to the rest of the plant. It is composed of long, thin cells called tracheids and vessel elements, which are arranged in a continuous tube-like structure. Secondary xylem is formed later in a plant's

DOI: 10.1079/9781836992288.0003

development and is responsible for the plant's overall structural support. It is made up of thick-walled cells called fibers and xylem parenchyma, which are arranged in a more diffuse and irregular pattern. In addition to its role in transporting water and minerals, xylem tissue also plays a role in plant defense. Some plants produce chemicals in their xylem tissue that can deter herbivores from feeding on them, or that can help to prevent the spread of diseases and parasites. Overall, xylem is an essential tissue in plants that is critical for their growth, development, and survival. Xylem tissue is essential for the proper functioning of a plant. It is responsible for providing the plant with the water and nutrients it needs to grow and survive. Without a functioning xylem system, a plant would quickly wilt and die.

When the electric current passes through these plant tissues, it disrupts the normal functioning of cells. The electrical energy causes chemical reactions and generates heat, ultimately damaging various cellular components, including the cell membranes, proteins, and enzymes. In overpowered applications, it is possible to see the plant tissues darken and get moist, as the contents of the xylem and phloem are expelled and exposed to the environment.

As both xylem and phloem are essential for the proper functioning of a plant and are critical for its overall health and growth, with the rupturing of the continuity of this essential flow, after a successful electrical weeding application, the plant is dead immediately, and should dry off after a few hours or days, depending on temperature.

The detailed investigation of how various plant morphologies respond to different electrical parameters in electrical weeding is a relatively new area of research, offering significant potential for deeper understanding through future studies.

Dicotyledons

Dicotyledons, also known as dicots, are a group of flowering plants that are characterized by having two cotyledons (seed leaves) when they germinate from a seed. Dicots are one of the two main groups of flowering plants, the other being monocots.

Some characteristics that are often found in dicots include: netted or reticulate venation in the leaves, which means the veins form a network pattern; stems with vascular bundles arranged in a ring; flowers with four or five petals and sepals, and a superior ovary (the ovary is located above the attachment point of the other flower parts); and fruit that is often a capsule, a drupe, or a berry. There are more than 200,000 known species of dicots, including many familiar plants such as roses, sunflowers, tomatoes, peppers, peas, beans, and oak trees. Dicots are important economically, as they include many crops and ornamental plants. They are also ecologically important, as they provide food and habitat for a wide variety of animals.

Dicots often have flowers that are bilaterally symmetrical, meaning that if you divide the flower down the middle, the two halves would be mirror images of each other. The leaves of dicots are usually broad and have a distinctive network of veins. The veins often form a pattern of loops or branches, and they may be more prominent on the upper surface of the leaf than on the lower surface. Many dicots have a distinctive type of structure which carries water and nutrients arranged in a ring around the outside of the stem. This arrangement is thought to be more efficient for the transport of water and nutrients than the arrangement found in monocots, which have a single, central vascular bundle.

Dicots are an important group of plants, and they are found in a wide range of habitats all over the world. They are especially diverse in tropical rainforests, where they make up most of the plant life. Some examples of important dicot families include the Rosaceae (which includes roses, apples, and almonds), the Fabaceae (which includes beans, peas, and lentils), and the Asteraceae (which includes sunflowers, daisies, and asters).

Dicotyledonous (dicot) plants typically have a taproot, instead of a fibrous root system, which means they have a single pivoting taproot, instead of a dense mass of thin, branching roots. Because of this characteristic, the fact that the roots and shoots derive from one single stem, if enough electrical energy is applied in a distributed manner throughout the general leaf area, all roots and branches coming out from the stem that connect to the remaining root tissue can be targeted successfully, leading to a high efficacy of the electrocution system, even if the remaining roots are still alive. So, because of this feature, *ceteris paribus*, dicots tend to be more sensitive (less energy is needed to perform a successful electrical weeding operation) than monocots.

Monocotyledons

Monocotyledons, also known as monocots, are a group of flowering plants characterized by having one cotyledon, or seed leaf, in their seeds. They are one of two main groups of angiosperms, the other being dicotyledons, or dicots. Some of the most well-known monocots include grasses, lilies, orchids, and palms.

Monocots can be distinguished from dicots by several characteristics. One key difference is the number of cotyledons present in the seed (Sabelli, 2012). Monocots have one cotyledon, while dicots have two. Monocots also tend to have parallel veins in their leaves, rather than the branched veins found in dicots. Monocot stem vascular tissue is also arranged in a circular pattern, rather than the more complex branching pattern seen in dicots.

Other characteristics that are often used to identify monocots include the presence of scattered vascular bundles in the stem, the presence of a single, large cambium ring in the stem, and the presence of scattered vascular bundles in the leaves. Monocots also tend to have flowers with parts arranged in threes, as opposed to the four- or five-part arrangement found in dicots (Raven *et al.*, 2005).

Despite these differences, monocots and dicots are not always easy to distinguish. Some plants may exhibit characteristics of both groups, making classification difficult. Additionally, some plants that were once classified as monocots have been reclassified as dicots based on new scientific evidence.

Overall, monocots are a diverse group of plants that play important roles in ecosystems around the world. They include important food crops such as rice, wheat, and corn, as well as many other economically and ecologically important species.

Typically, monocots have a fibrous root system, in which the roots branch out in many directions from the base of the plant, rather than forming a deep taproot. These roots are generally thin and often grow close to the surface of the soil. The fibrous root system allows monocots to quickly and efficiently absorb water and nutrients from the soil, which is important for the plant's growth and development (Smith *et al.*, 1940).

One characteristic that sets monocots apart from dicots is the presence of adventitious roots. Adventitious roots form in unusual places on the plant, such as on the stem or leaves. They are often used by monocots to anchor the plant in the soil or to absorb water and nutrients from the surrounding environment.

In addition to the fibrous root system, many monocots also have specialized structures called rhizomes, which are underground stems that can produce new roots and shoots. Rhizomes allow monocots to spread and reproduce vegetatively, rather than relying on seeds for reproduction. This allows them to colonize new areas and establish themselves in a variety of environments.

Overall, the root systems of monocots are adapted for efficient absorption of water and nutrients, and for the ability to spread and reproduce vegetatively. These adaptations allow monocots to thrive in a wide range of environments and play important roles in many ecosystems around the world.

This fibrous root system is well-suited to monocots because it allows them to take up water and nutrients from a wide area, which is important in environments where the soil is not particularly fertile or where there is not a lot of water available. Fibrous root systems are also generally more resistant to drought and erosion than taproot systems, because they have a larger surface area for absorbing water and because they are less likely to be pulled out of the ground when the soil is disturbed.

Due to this sparse root structure without a pivoting central unification, more energy is needed for high-efficacy electrical control, since there are underground structures that may survive a lower energy application and regrow, or even reestablish surviving xylem and phloem systems to keep the remaining tissues of the plant alive. So, because of this feature, *ceteris paribus*, monocots tend to be less sensitive (more energy is needed to perform a successful electrical weeding operation) than dicots.

The different arrangement of the vascular bundles also causes different efficacy of electrical weeding between monocots and dicots due to effects such as skin effect. In electromagnetism, skin effect is the tendency of an alternating electric current (AC) to become distributed within a conductor such that the current density is largest near the surface of the conductor and decreases exponentially with greater depths in the conductor.

Mistakes from Other Studies on Dicot vs Monocot Regarding Lethal Energy Threshold

The following is the author's proposed perspective on lethal energy threshold regarding electrical weeding, based on practical experience throughout years of field results, and there is space for more formal studies to be performed for validation (or not).

Several past studies on electrical weeding have incorrectly concluded that dicotyledonous plants with pivoting (tap) roots exhibit a diminished response to electrical treatment compared to monocots with fibrous root systems. The primary reasoning behind this misconception has been the assumption that dicots, particularly those with higher lignin content, are more resistant to electrical current due to lignin's potential insulating properties when dry. However, empirical evidence contradicts this assumption, demonstrating that dicots with taproots are actually more susceptible to electrical weeding than their monocot counterparts.

The root cause of this error lies in the experimental methodology used in these studies. Many relied on constant voltage application rather than constant power application. Under a constant voltage system, the electrical energy delivered to a plant is largely influenced by its electrical impedance. Plants with higher lignin content tend to have greater dielectric properties and higher electrical resistance, resulting in lower current flow and, consequently, reduced power delivery. This effect creates the illusion that such plants are more resistant to electrical weeding.

However, when a constant power system is employed—where power delivery is dynamically adjusted to ensure a consistent energy transfer regardless of impedance—an entirely different outcome emerges. Empirical tests show that dicots with pivoting roots are significantly more vulnerable to electrical weeding than monocots with fibrous root systems, which generally have a higher total root volume. The constant power approach ensures that energy is efficiently transmitted into the plant's vascular system, particularly through the xylem and phloem, where the current is most effective in causing lethal damage.

This finding aligns with the latest technological advancements in electrical weeding, where state-of-the-art power electronics prioritize semiconstant power regulation over traditional constant voltage methods. Such advanced control strategies optimize energy delivery for maximum efficacy, disproving the outdated assumption that lignin-rich plants are inherently more resistant. In reality, the anatomical structure of dicots—with their centralized vascular bundles—facilitates a more efficient conduction of electrical current into deeper plant tissues, leading to a more effective lethal dose.

In summary, past mistakes in electrical weeding research stemmed from a flawed voltage-based experimental design that failed to account for impedance effects. When using power-regulated electrical weeding systems, dicots with pivoting roots are actually easier to eliminate than monocots with more dispersed root structures. As electrical weeding technology continues to evolve, it is crucial that future studies adopt constant power methodologies to ensure accurate assessments of plant susceptibility.

References

Raven, P.H., Evert, R.F. and Eichhorn, S.E. (2005) *Biology of Plants, 6th edn.* W.H. Freeman, New York, USA.

Sabelli, P.A. (2012) *Seed Development: A Comparative Overview on Biology of Morphology, Physiology, and Biochemistry Between Monocot and Dicot Plants*. Springer, Dordrecht, Netherlands.

Smith, G.M., Overton, J.B., Gilbert, E.M., Denniston, R.H., Bryan, G.S. and Allen, C.E. (1940) *A Textbook of General Botany, 3rd edn*. Macmillan Company, New York, USA.

4

How Plants Die from Electrocution

Principles

The process of electrical plant control requires that each target plant become part of the electrical circuit. One of the recent innovations of this technology is based on the fast dynamic adjustment of the ratio of voltage and current that relates to the electrical impedance of the system, ensuring the semiconstant power throughout application, providing a more homogeneous power output. The proper use of applicator and electrode designs minimizes the electricity applied to bare soil, consequently maximizing energy applied into plants. This also implies that the impact of electricity to the soil is limited.

The degree of electrical energy consumed in a plant to ensure tissue damage in the shoot and root of the plant can be used as a measure of the energy required for a successful and lasting electrical application result. A proper level of electrical energy applied to individual plant and root systems is crucial to prevent recovery or regrowth. Therefore, at the plant level, the effectiveness of tissue destruction by electrical power can be considered binary: the plant is dead, or it is not dead. The amount of energy required mainly varies depending on plant species (including its morphology, root volume, and age), soil type, and soil moisture conditions.

At the cell level, the destruction of both stem and root cells due to electrical flow can be observed. Measurements of the plant electrical impedances indicate changes of plant tissue electrical conductivity after electrical application (Zhang and Willison, 1993). Cell walls are punctured by the application of voltage, and, over time, as the current flow is increased, tissue damage starts to manifest. The initial and final current which passes through the plant tissue can be distinguished. Its destructive effect has been described as a heating of cell fluids, expanding into a vaporous phase which results in the irreversible rupture of cell walls and membranes (Dykes, 1979). Additionally, the vascular bundle cells burst irreparably, interrupting the plant's water and nutrient transport, resulting in a breakdown of metabolic processes.

In general, the mode of action of electrical plant destruction is comparable to the action of a chemical herbicide due to the high-voltage direct current (DC) running through the entire plant and roots, which enables additional control of root runners, and thus prevents vegetative reproduction. The application dosage of chemical vegetation control products is defined by the application rate in liters per hectare (l ha^{-1}). Likewise, for the Electroherb™ technology, the treatment

DOI: 10.1079/9781836992288.0004

intensity can be defined in kilowatt hours per hectare (kWh ha^{-1}), which is affected by the working speed in kilometers per hour (km h^{-1}), the power output in kilowatts (kW), and the geometry of the electrodes.

Plant Electrocution Resistive Circuit

A worthwhile objective is to mathematically define the electrical variables of a perfect plant electrocuting system and their interactions. With this mathematical description, the relations between the resistances of the system, and how different soils and plants might affect the efficacy of the system, can be further understood. Conclusions should be able to further explain and determine factors of influence on the system's efficacy.

In some cases, a single applicator might touch more than one plant at the same time. In those cases, energy consumed will be divided through all plants touched by the applicator. Equation 4.1 defines total resistance for parallel resistances as:

$$\frac{1}{Rp_t} = \Sigma \frac{1}{Rp_n}$$

Equation 4.1. Total resistance for parallel resistances.

The plant electrocution resistive circuit is formed by the soil resistance summed together with the parallel resistances of plants touched by a single applicator (Fig. 4.1).

In an ideal plant electrocution system:

- The electric tension (voltage) of the system is defined by the secondary of the transformer.
- The system delivers maximum power at all times.
- The electric tension (voltage) adjusts itself immediately to ensure it operates at maximum power.
- Cos φ = 1

This means that, in this ideal system (Eqn 4.2):

$$P_t = V * I$$

Equation 4.2. Ohm's Law.

Individual Plant Energy Consumption in a System of Plants

Through internal lab tests, it was observed that the lethal amount of energy for an average 0.15 m broad-leaved plant usually varies between 100–1,000 J (Almeida, 1988). In a "perfect" plant electrocution, the equipment will adjust its tension voltage (V) every moment for the varying total resistance (R_t) to ensure that, with a perfect Cos φ = 1, its power consumption equals its total power capacity at all times. This considers that, although maximum power generation is limited, "perfect" equipment can speed up, or even change, the broadness of application, so the energy per plant is at the ideal value at all times, while maintaining maximum power.

Given the power available for plant electrocution is total power minus power consumed by the soil resistance (Eqn 4.3):

$$P_{tp} = P_t - R_s I_t^2$$

Equation 4.3. Power available for plant electrocution.

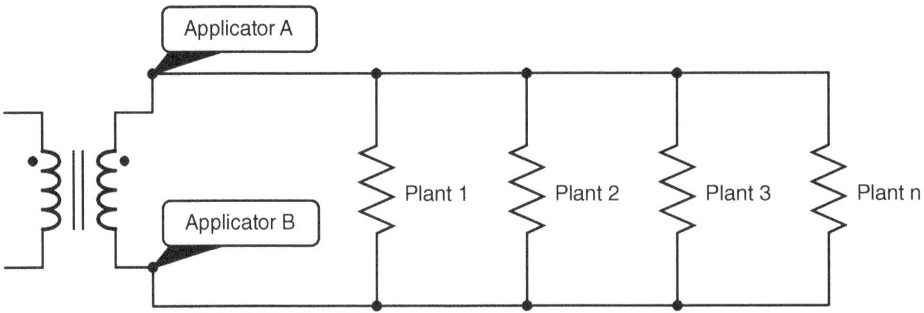

Fig. 4.1. The plant electrocution resistive circuit.

And that the resulting power must be consumed either by plant n, or by all the other plants except plant n (Eqn 4.4):

$$P_{tp} = R_n I_n^2 + R_{Eq_{t-n}} I_{t-n}^2$$

Equation 4.4. Power available for plant electrocution "n."

Therefore, the power available to kill any specific plant is determined as total power minus power lost in the soil, minus power used to electrocute other plants (Eqn 4.5).

$$R_n I_n^2 = P_t - R_s I_t^2 - R_{Eq_{t-n}} I_{t-n}^2$$

Equation 4.5. Total power minus power lost in the soil, minus power used to electrocute other plants.

According to Ohm's Law, that can also be viewed as (Eqn 4.6):

$$V_p^2 / R_n = V^2 / R_t - V_s^2 / R_s - V_p^2 / R_{t-n}$$

Equation 4.6. Ohm's Law view of total power minus power lost in the soil, minus power used to electrocute other plants.

We can conclude that energy consumption by the plant n is:

- proportional to total power: $\left(\frac{V^2}{R_t}\right)$
- proportional to time of exposure: t
- inversely proportional to velocity of application, which implies a proportional reduction of time of exposure.

So, we can further conclude that, *ceteris paribus*:

- Higher soil resistances diminish efficacy (there is much more to be said on this regard, since soil resistance is related to the space between electrodes, and a higher soil resistance may increase the depth of current penetration into the roots, but this is to be explored further in the next chapters).
- Higher plant n resistances diminish efficacy.
- Higher resistances on the plant system, excluding plant n, increase efficacy.

One other practical conclusion:

- A system with plants with similar electrical resistances and similar air/root systems will have a more evenly distributed efficacy.

Plant Electrocution: Individual Plant Efficacy Parameters

Root's electrical sensitivity constant

Ideally, the efficacy of plant electrocution can be considered to be binary (plant is dead or not dead). Of course, in many cases, and especially in well-developed grasses with vegetative propagation, it is possible that only part of a plant is controlled, while the other part of a plant remains unaffected.

Efficacy is directly affected by the relation between the fatal energy consumption per volume of root needed at the individual plant's root and the volume of the root system. This relation might vary a little from plant to plant and due to conditions (temperature, humidity, etc.), but for set conditions, the electrical sensitivity constant for a specific plant morphology can be determined as (Eqn 4.7):

$$C(n) = \frac{E(nroot)}{Ap(nroot)}$$

Equation 4.7. Root's electrical sensitivity constant.

- *C(n)* is the electrical sensitivity constant of that specific plant, in determined conditions.
- *E (n root)* is the energy consumption at the n plant root.
- *Ap (n root)* is the volume of the n plant's root system.

Plants can be seen as serial resistances

For general weeding purposes, the success of electrical weeding is fundamentally tied to its ability to disrupt the root system rather than merely damaging the aerial (aboveground) portion of the plant. This is not the case only in some specific siccation cases, such as killing the air system of some plants for maturing grains all at once for harvesting. That said, while defoliation can cause temporary setbacks in plant growth, many species, particularly perennials, are capable of regrowth if their root system remains intact. In contrast, a plant will inevitably die if its root system is sufficiently impaired, as it will be unable to transport essential nutrients and water to sustain metabolic functions.

Physiological justification for root system targeting

Plants rely on their vascular system—comprising xylem and phloem—to transport water, nutrients, and organic compounds throughout the organism. The root system is responsible for:

- **water uptake** from the soil through osmosis;
- **nutrient absorption** (e.g. nitrogen, phosphorus, potassium);
- **anchoring the plant** and providing stability.

If electrical energy effectively damages the root system, two key failures occur:

- **Impairment of water and nutrient transport.** Without a functional xylem and phloem, water cannot reach the aerial tissues, leading to desiccation and metabolic collapse.
- **Destruction of meristematic tissues.** Many plants regenerate from their root apical meristems, and their destruction prevents regrowth.

Conversely, if only the aboveground portion of a plant is affected, the plant may survive by sprouting new shoots from its root system.

Total plant resistance: The sum of air and root resistance

The electrical impedance of a plant, denoted as total plant resistance (*Rn*) (Fig. 4.2), is the combined resistance of:

- ***Rn_air.*** The resistance of the aerial system, including leaves and stems.
- ***Rn_root.*** The resistance of the root system.

Mathematically:

$$Rn = Rn\_air + Rn\_root$$

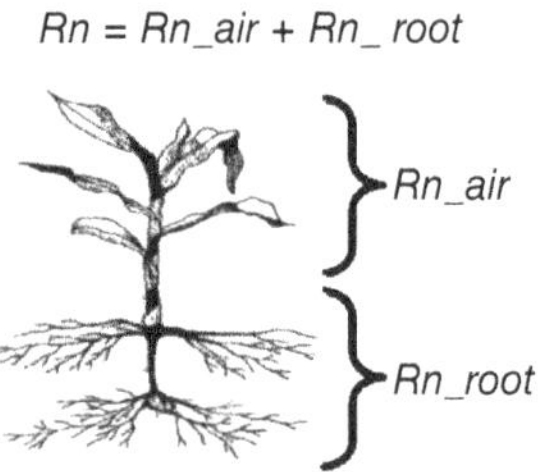

Fig. 4.2. Total plant resistance schematic.

Implications for electrical weeding strategies

- **Optimizing soil conductivity.** Since root damage is essential, ensuring proper soil conductivity (e.g. through moisture control for application at optimized parameters) can improve electrical energy transmission to roots and minimize losses.
- **Maximizing root penetration.** Applying higher energy parameters or using optimized electrode configurations can direct more current through the root zone.
- **Adjusting electrical parameters.** Parameters may be tuned to overcome root resistance and maximize root system electrocution.

Individual plant energy consumption

Considering total plant resistance (*Rn*) as the sum of two resistances, *Rn_air* and *Rn_root*, as a direct consequence of Ohm's Law, the amount of current (I) flowing through the whole plant will be, where voltage is V, V/*Rn* (Fig. 4.3, middle triangle).

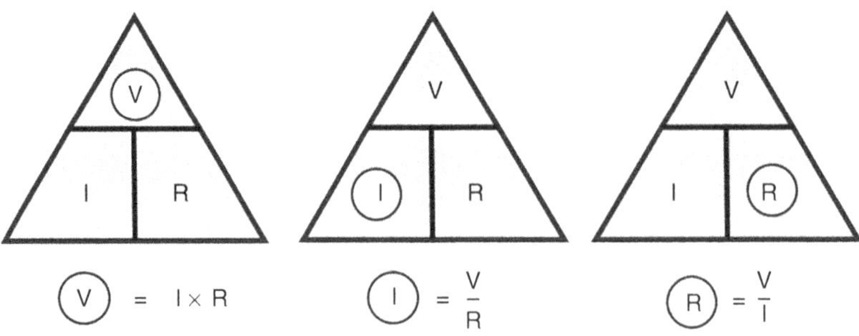

Fig 4.3. Ohm's Law schematically.

As power is given by Ohm's law, and energy (E) is power (P) applied in time, given by (Eqn 4.8):

$$E = P * t$$

Equation 4.8. Energy (E) is power (P) applied in time.

Therefore, given a set voltage (V; considering a voltage source, which may not be the best kind of power source to control weeds) and a set amount of time applying this voltage, a lower total resistance (Rn) means more energy flow through the whole plant, according to Eqn 4.9 below:

$$E = \frac{V^2 * t}{R_n}$$

Equation 4.9. Total resistance increases are inversely proportional to energy consumption.

Individual plant sensitivity due to air system volume/root system volume ratio

As total resistance (Rn) is the sum of the air system resistance and the root system resistance (Eqn 4.10)

$$Rn = Rn\,air + Rn\,root$$

Equation 4.10. Total resistance (Rn) is the sum of the air system resistance and the root system resistance.

Given a set total plant resistance (Rn), and a set energy applied, as the goal is to kill the root system of the plant and not the air system, the energy consumed at the root system (Er) is influenced by the energy consumed at the air system (Ea). Therefore, total energy consumed at the root system is (Eqn 4.11):

$$E = Ea + Er = \frac{V_{air}^2 * t}{R_{n\,air}} + \frac{V_{root}^2 * t}{R_{n\,root}}$$

$$Er = E - \frac{V_{air}^2 * t}{R_{n\,air}}$$

$$Er = E - \frac{I^2 * R_{n\,air}^2 * t}{R_{n\,air}}$$

$$Er = E - I^2 * R_{n\,air} * t$$

Equation 4.11. Total energy consumed at the root system where the time applied (t) and total energy flow (E) are set.

Because of this, a lower air system resistance ($R_{n\,air}$) means more energy being consumed at the roots and consequently a higher individual plant sensitivity to electrical weeding.

Considering the plant is dead if you kill the roots independently of what happens to the air system:

- For a given total resistance (Rn), a higher root system resistance means lower air system resistance.

Therefore (Fig. 4.4):

- More leaves mean more contact area, therefore more leaves mean less resistance.
- More roots mean more contact area, therefore more roots mean less resistance.

Therefore, plant sensitivity to electrical weeding can be represented as (Fig. 4.5):

When compared with broad-leaved plants, grasses usually have lower air/root systems (and consequently higher air/root system electrical resistances), and, in general, grasses have lower sensitivities to electrical weeding when compared with broad-leaved plants.

Video weed type identification

The plants can be identified by video recognition for targeted application. This ensures that only

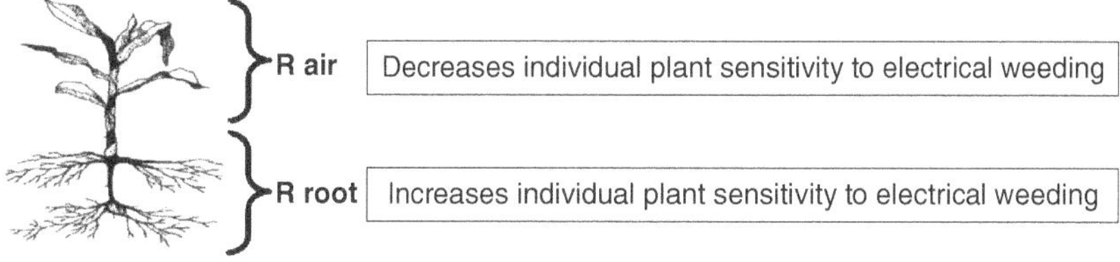

Fig. 4.4. Total plant resistance schematic with sensitivity parameters.

the desired plants are treated and thus reduces the risk of damage to persons or animals present. This requires at least one appropriately equipped camera with evaluation electronics, which is permanently directed toward the ground during operation. If a modular design is used for high-voltage generation, the individual modules can be switched on or off depending on the plant presence. For additional risk minimization when directly touching an electrode, at least one movable electrode can be used with low vegetation density. This electrode moves to the target plant, transmits current to the plant for a sufficiently short time and is then switched off again.

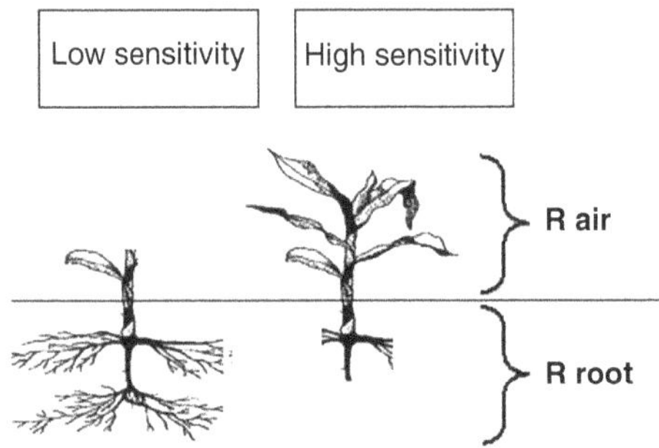

Fig. 4.5. Total plant resistance schematic with sensitivity parameters.

One way to increase the efficiency of the overall system is to analyze the density of vegetation. With the help of plant identification, this and the types of plant can be determined. If it is known which energy the plants need at that time for a successful treatment, the energy can be adjusted by the driving speed depending on the vegetation. This concept is called speed control (SC). For example, the driving speed is higher at a high vegetation density. With the help of this control, the average operating time per area can be reduced (Fig. 4.6). Furthermore, the risk of fire is reduced if the driving speed is kept high.

Soil Moisture Impacts Efficiency

Catherine P.D. Borger and Miranda J. Slaven from the Australian Department of Primary Industries and Regional Development produced

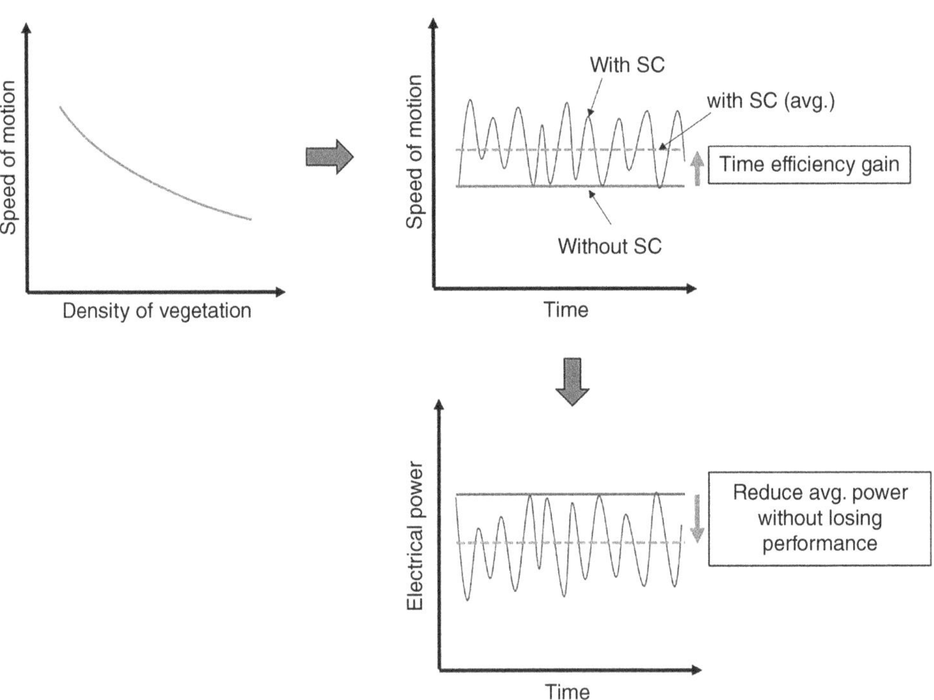

Fig. 4.6. Speed control (SC) based on video identification leads to time efficiency gain and reduction of average power without losing treatment quality.

a paper regarding the issue of soil moisture impacts on efficiency. The findings are coherent with the expected results. The following is a depiction of their findings (Borger and Slaven, 2024).

Depending on application parameters, the effectiveness of electric weed control (EWC) can be tied to environmental conditions, particularly soil moisture levels. To explore this relationship, two field trials were conducted, each evaluating the impact of soil moisture variations on weed management efficiency using EWC technology. These experiments, conducted in 2022 and 2023, aimed to quantify the influence of moisture content on the power delivery and weed suppression effectiveness of the system.

For both trials, EWC was applied using a New Holland TS100A tractor fitted with a rear-mounted XPower power supply unit. A front-mounted XPU applicator, spanning 1.2 m in width and equipped with three arrays of electrodes, delivered the electric current. This system, designed by CNH Industrial Australia Pty Ltd, was capable of delivering 36 kW of power, distributed among 12 inverters, each with a maximum output of 3 kW.

The first field trial took place at the Department of Primary Industries and Regional Development (DPIRD) Northam Research Station (coordinates −31.6532, 116.6951). The soil at this location was classified as brown sand, containing 5% gravel, with a conductivity of 0.037 dS m^{-1} and a pH($CaCl_2$) of 4.4.

The experimental design targeted three weed species:

- Annual ryegrass (*Lolium rigidum* Gaud.)
- Wild radish (*Raphanus raphanistrum* L.)
- Kikuyu grass (*Cenchrus clandestinus* Hochst. ex Chiov.)

Each species was sown at a density of 200 seeds per square meter. To simulate varying soil moisture conditions, three levels of artificial rainfall were applied immediately before treatment:

- 0 mm (dry conditions)
- 2 mm (moderate moisture)
- 3 mm (high moisture)

The trial followed a split-plot design, with EWC as the primary treatment and weed species by soil moisture as the secondary factor, randomized within subplots (2 m × 4 m), with three replications. The weeds were treated at different growth stages: annual ryegrass (late tillering), wild radish (anthesis), and kikuyu grass (early tillering).

On August 12, 2022, EWC was applied at a speed of 3 km h^{-1}. The power drawn during treatment fluctuated between 1,155–3,053 W per inverter per second, depending on plant density. Since kikuyu plants were smaller and more sparsely distributed compared to wild radish, power demand varied accordingly. The system's power draw was directly influenced by the level of plant contact, meaning that areas with lower weed biomass or density required less power.

The second trial was conducted at the DPIRD Wongan Hills Research Station (coordinates −30.8465, 116.7409). Unlike the previous site, this location had yellow-grey sandy loam soil with 15–20% gravel content, a higher conductivity of 0.082 dS m^{-1}, and a pH($CaCl_2$) of 6.5.

For this experiment, annual ryegrass (*Lolium rigidum* Gaud.) was sown at the same density of 200 seeds per square meter. The simulated rainfall treatments were adjusted to represent a broader range of soil moisture conditions:

- 0 mm (dry conditions)
- 10 mm (moderate moisture)
- 20 mm (high moisture)

The split-plot design was maintained, with six replications per treatment, and slightly larger plots (2 m × 5 m) to improve statistical reliability. The treatment took place on August 23, 2023, when annual ryegrass was at the late tillering stage. Unlike the previous year, the application speed was reduced to 2 km h^{-1}, allowing for a higher dose of electric current per plant.

During this trial, the average power draw per inverter was 2,517 W per second, and, unlike the Northam experiment, power output remained consistent across treatments. This was largely attributed to the uniform plant size and density, which reduced variability in the system's energy requirements.

To quantify the relationship between soil moisture and treatment efficacy, several parameters were measured:

- Soil moisture was recorded to a depth of 12 cm using a HydroSense II Handheld

Soil Moisture Sensor (Campbell Scientific Australia). In 2022, one measurement per plot was taken, while in 2023, two measurements per plot were recorded for improved accuracy.

- Soil characteristics were assessed using methods standardized by CSBP Ltd. (2013).
- Weed density and biomass were measured three weeks posttreatment. Two quadrats (50 cm × 50 cm) were sampled per plot.
- Biomass samples were dried at 60°C for three days, after which their dry weight was recorded.
- Statistical analysis was conducted using ANOVA, with least significant difference (LSD) tests applied to compare treatment means. Residual plots were examined to confirm the normal distribution of the data, and transformations were applied where necessary to ensure statistical validity.

To ensure accurate soil moisture readings, the HydroSense probe was calibrated against gravimetric soil moisture measurements at both trial sites. A calibration zone was established near each site, where ten 20-liter buckets with drainage holes were filled with water and placed at 10-meter intervals. As the water gradually infiltrated the soil, measurements were taken at five time points over 48 hours.

Two methods were used for comparison:

- HydroSense probe readings were taken at each time point.
- Soil core samples (4 cm diameter × 10 cm depth) were extracted, dried at 105°C for three days, and used to calculate volumetric water content based on bulk density.

The HydroSense probe consistently underestimated soil moisture:

- At Northam (2022), readings were 4.75% lower than the soil core method.
- At Wongan Hills (2023), readings were 5.75% lower.

These discrepancies were accounted for by applying *correction factors* to field moisture measurements to improve accuracy. The volumetric soil moisture (with standard error) of the soil at each site at field capacity and then 48 h after reaching field capacity, following assessment using a HydroSense probe or soil cores.

In the 2022 trial, EWC significantly reduced weed density and aboveground biomass across all species compared to untreated plots. Among the species studied, wild radish exhibited the highest biomass, highlighting species-specific variability in response to treatment (Table 4.1).

However, the effect of soil moisture on weed density and biomass was not statistically significant (Table 4.2). The volumetric soil moisture content in the experimental plots increased only slightly, from 21% to 25%, which may explain the lack of a strong moisture effect. This limited variation was attributed to higher-than-average rainfall during the growing season—385 mm from April to October, compared to the long-term average of 359 mm. As a result, opportunities to conduct treatments in dry soil conditions were minimal, reducing the ability to assess the full impact of soil moisture on EWC.

Although the results showed a nonsignificant trend, weed density and biomass were slightly lower in plots without additional rainfall (0 mm) compared to those that received 3 mm rainfall before treatment.

In contrast to the previous year, the 2023 trial demonstrated a strong relationship between soil moisture and annual ryegrass survival. The density and biomass of ryegrass increased significantly

Table 4.1. Species-specific variability in response to treatment. (Table used with permission from Slaven and Borger.)

Weed control	Weed species	Density (plants m^{-2})	Biomass (g m^{-2})
None	Kikuyu	98 (9.89)	21 (1.34)
	Ryegrass	79 (8.89)	105 (2.03)
	Wild radish	64 (7.98)	305 (2.49)
Electric weed control (EWC)	Kikuyu	2 (1.45)	1 (0.21)
	Ryegrass	51 (7.15)	30 (1.51)
	Wild radish	35 (5.92)	129 (2.23)
p-value (least significant difference)		<0.001 (1.56)	0.018 (0.43)

Table 4.2. Soil moisture on weed density and biomass. (Table used with permission from Slaven and Borger.)

Weed control	Rainfall treatment (mm)	Soil moisture (%)	Density (plants m^{-2})	Biomass (g m^{-2})
None	0	8.4 (±0.8)	94 (9.68)	244
	10	12.3 (±0.1)	84 (9.14)	262
	20	15.9 (±1.7)	95 (9.74)	274
Electric weed control (EWC)	0	9.2 (±0.8)	43 (6.58)	68
	10	14.2 (±1.1)	54 (7.35)	141
	20	16.2 (±1.3)	77 (8.79)	200
p-value (least significant difference)			0.001 (1.730)	<0.001 (64.80)

with higher soil moisture, particularly following 20 mm of simulated rainfall. The volumetric soil moisture increased from 9% to 16%, confirming that higher moisture levels reduced the effectiveness of EWC.

These findings align with manufacturer recommendations, which caution against applying EWC immediately after rainfall, as moist plants can reduce treatment efficacy. The data further reinforces the assumption that higher soil moisture increases electrical dissipation into the soil, thereby decreasing the lethal effect on plant tissues (Slaven and Borger, 2023).

Notably, the weeds in both the 2022 and 2023 trials were mature, and the ryegrass population in 2023 was particularly dense. Given the experimental setup, complete weed control was not expected, as treatment speeds of 2–3 km h^{-1} were selected—below the recommended 1 km h^{-1} for grasses and 2 km h^{-1} for broad-leaved species. This deliberate choice allowed researchers to better observe the impact of soil moisture on treatment efficacy, as full control would have masked any moisture-related effects.

The observed impact of soil moisture aligns with industry assumptions that higher moisture reduces soil resistance, allowing electric current to disperse away from plant tissues (Slaven and Borger, 2023). However, an alternative explanation suggests that plants experiencing drought stress may be less capable of recovering from electric treatment, similar to trends seen in mechanical weed management (Slaven and Borger, 2023).

The results of the 2022 and 2023 experiments demonstrate that soil moisture significantly affects the efficacy of EWC, with higher moisture levels reducing its effectiveness. The findings reinforce manufacturer recommendations to avoid treatment in wet conditions and suggest that drier conditions may enhance weed suppression by limiting plant recovery.

Conclusion

The work of Borger and Slaven is coherent with the author's field experience, which suggests that a higher moisture content in the system may have two distinct effects that may have a detrimental effect on efficacy, although in a "perfect" application, lower soil resistance may mean less losses in the soil, which has a positive effect in electrical weeding energy efficiency:

- Plants with hydric stress are more sensitive and will be controlled at a lower lethal energy threshold.
- Lower soil electrical resistance will increase energy losses related to parallel currents through the soil that do not go through the plants, in cases where the electrodes are not height-controlled. This happens when the electrodes accidentally or per design touch open ground, creating a parallel soil–soil circuit that has no weeding effect.

How Electrode Distance Affects Average Current Depth

Effective electrical weeding depends on the precise delivery of electric current deep enough into the soil to destroy not only the aerial parts of plants but critically, their root systems. Soil acts as a crucial part of the electrical circuit. Its electrical resistance determines how much current actually passes through the roots, how deep the current can penetrate, and how much energy is lost to the ground instead of affecting the plants.

Low soil resistance (e.g. in humid soils) may prevent current from reaching deep roots unless electrode distance or disposition, and applied voltage, are optimized. This is simply because lower soil resistance will have the effect of higher energy consumption at the top soil level, when compared to the deeper layers, since energy reaching the deeper layers will have to go through more root length. On the other hand, dry and less conductive soils allow deeper and more efficient energy distribution but increase energy losses through the Joule effect in the soil. Thus, the system must reduce electrode distance if soil impedance is high, concentrating energy between electrodes, and increase electrode distance if soil impedance is low.

This effect becomes clear when you model a plant–root and soil resistive system (Fig. 4.7).

Considering a nonzero root resistance (hereby represented by R1 and R2), even if the soil resistance is homogeneous throughout the depth layers (hereby represented by R3, R4, and R5), the lower the soil resistance, the lower the current going through the deeper roots, represented by R3. To create an intuition, this becomes clear when using a *reductio ad absurdum* where the soil resistance is zero (R3 = R4 = R5 = 0) and the root resistance is not zero (or, at least R1 = /0)—a scenario where all current would flow through the zero electrical resistance of the very top soil layer (R3), and where, consequently, zero current would reach the roots.

Thereby, as soil resistance increases (considering homogeneous soil resistance throughout the different soil layers), the deeper the average current goes in the plant roots (Fig. 4.8).

Thereby, the practical implication is that for any nonzero soil resistance (considering homogeneous soil resistance throughout the different soil layers), the further apart the electrodes are, the higher the soil resistance, and the deeper the average current goes into the roots (Fig. 4.9).

The downside of placing electrodes further apart is to increase soil resistance and, therefore, the overall resistance of the system. This means two different detrimental effects happen as electrodes are placed further apart—higher overall resistance means overall voltage for the same power consumed is increased, and higher soil resistance means more energy is consumed in the soil and not in the plant.

Since shallower roots allow for operating with electrodes closer together, which, in turn, allows for lower energy losses throughout the soil, operating when the targeted plants have shallower roots ensures higher operational energy efficiency. Added bonuses relate to the fact that with lower soil resistance, constant power equipment may operate at a lower voltage level—which also means a safer operation.

The conclusion is that killing weeds when they are still small is not only more effective because they are inherently easier to kill, since

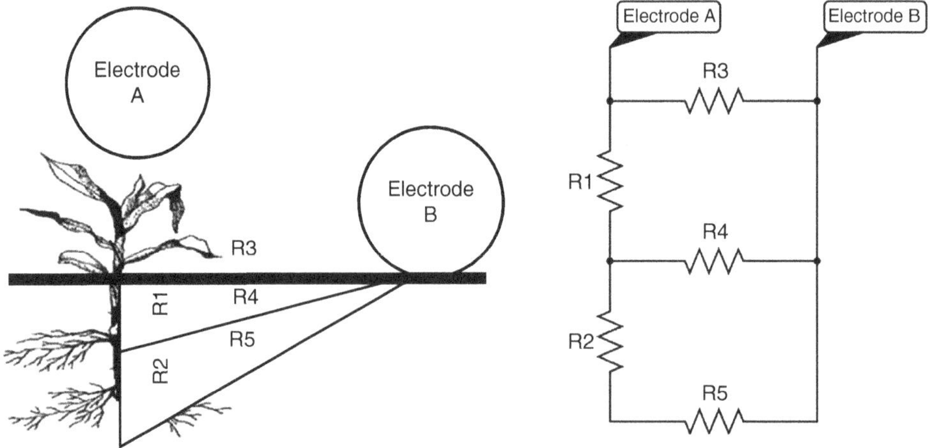

Fig. 4.7. Plant resistive system model in relation to root depth.

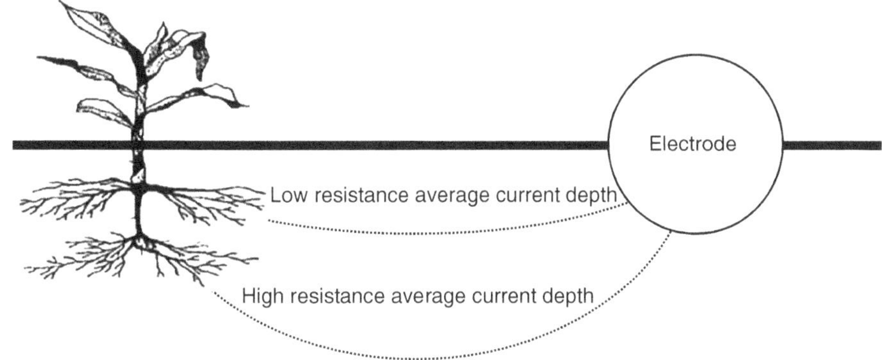

Fig. 4.8. Average current depth for the same electrode distance and different soil resistance.

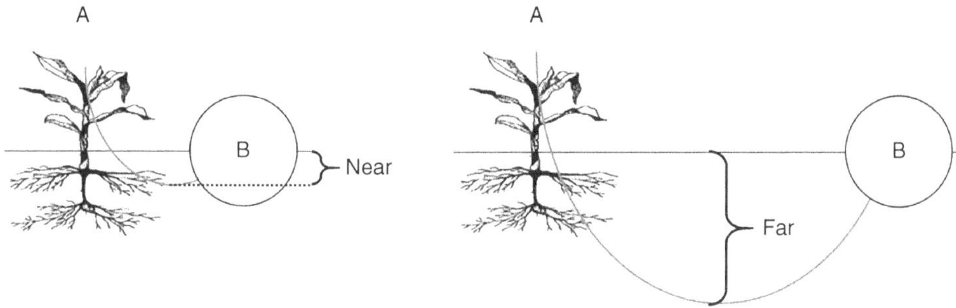

Fig. 4.9. Average current depth for the same soil and different electrode distances.

you will need proportionally less energy, but also because you can operate in such a way that ensures less energy will be lost heating up the soil.

Mowing before Electrical Weeding

The shielding problem: When taller plants protect smaller ones

In fields where weed biomass is dense and layered, especially in late-stage infestations, it is common to observe that larger, dominant plants physically shield smaller plants beneath them. These canopies prevent the smaller weeds from making contact with the electrode surfaces, and consequently from receiving sufficient current to die as a result of the treatment. When weeds beneath the canopy avoid contact, no current flows through them, they remain alive, and may resprout aggressively after taller plants are killed. Therefore, this creates a false visual efficacy, where treatment seems successful at first but regrowth occurs shortly after.

By mowing before application, operators can flatten or open the canopy, improve electrode access to smaller or hidden plants, and expose root systems that were previously unreachable, therefore increasing uniformity of treatment across all weed sizes.

In conclusion, in multilayered or competitive weed populations, mowing can significantly enhance electrical weeding performance by ensuring even and complete exposure.

The umbrella paradox: When mowing reduces root connectivity

In contrast, when large monocots or deep-rooted perennial species (e.g. *Sorghum halepense*, *Cynodon*, or *Imperata cylindrica*) are present, mowing may have unintended side effects. These plants rely on tall stems and dense vascular bundles that serve as direct, conductive channels to their deep root systems. Mowing removes part of this conductive tissue, so the remaining plant may no longer provide an electrical pathway to the root zone. As a result, the electrical current only affects the cut surface or upper section, not the crown or rhizomes. Some root systems remain viable and regrow vigorously, sometimes with enhanced resistance due to partial stress responses.

In deep-rooted monocots, mowing can decrease root lethality. Regrowth may be faster and more aggressive if the underground biomass survives, where visual dieback is often temporary, followed by robust regrowth.

Given these contrasting outcomes, the decision to mow should be *situational*. Only mow before electrical weeding when:

- the weed canopy is layered and dense;
- there is significant shading or shielding of smaller weeds;
- the goal is to maximize first-pass coverage;
- you expect to have a second pass to kill remaining weeds that may appear from regrowth;
- mowing is never done aggressively, since this may increase the severity of the issue relating to no circuit available into the surviving root systems.

From a biophysical perspective, the effectiveness of electrical weeding relies on contact area between electrode and plant tissue, continuity of vascular paths for current flow, and enough current density reaching the crown or root system.

Conclusion

Mowing before electrical weeding is not universally beneficial. While it helps improve contact uniformity and reach in multilayered or high-density broadleaf infestations, it can be counterproductive when dealing with robust monocots, whose root system control through electrical weeding relies on uninterrupted stem conductivity for electrical penetration.

Understanding the architecture and physiology of the weed population is crucial in deciding whether to mow. A species-specific, condition-driven strategy offers the highest probability of successful root-kill (systemic control) with minimal regrowth.

References

Almeida, F.M. (1988) *O uso de descarga elétrica no controle de plantas daninhas*. Universidade Estadual Paulista (UNESP), Botucatu, Brazil.

Borger, C.P.D. and Slaven, M.J. (2024) *Does Soil Moisture Impact Electric Weed Control Efficiency?* Department of Primary Industries and Regional Development (DPIRD), Northam, WA, Australia.

Dykes, W.G. (1979) *Plant Destruction Using Electricity. U.S. Patent 4,177,603*. Lasco, Inc., Vicksburg, MS, USA.

Slaven, M.J. and Borger, C.P.D. (2023) *What is the Best Fit for Electric Weed Control in Australia?* Department of Primary Industries and Regional Development (DPIRD), Northam, WA, Australia.

Zhang, M.I.N. and Willison, J.H.M. (1993) Electrical impedance analysis in plant tissues: impedance measurement in leaves. *Journal of Experimental Botany* 44(8): 1369–1375.

5

Application Possibilities and Electrode Arrangements

Spark Discharge and Height Selectivity

As an alternative to the direct contact application practice, researchers reported findings on electrical spark discharge operation by high-voltage discharges from an energy storage capacitor (Savchuk and Bayev, 1975). This can also be done with electrodes connected to alternating current (AC) or direct current (DC) voltage. Sparks from the charged electrodes enter the plant when the plant passes near the electrodes and initiates a voltage breakdown (Diprose and Benson, 1984). In 1981 (Wilson and Anderson, 1981), investigations were reported for an electrical discharge system from an American company (Lasco, "Lightning Weeder"). This method poses safety issues once the electrodes are exposed at a height, and may not be a good fit for applications that do not relate specifically to height selectiveness. That said, it has been successfully used at a commercial level for height selectivity weed management. An example is the aforementioned patent from Lasco (Dykes, 1979).

Direct Contact Electrodes

Many of the inventions followed the concept of physical contact by electrodes directed to the plant to establish the electrical circuit. In the same way, Electroherb™ technology only works on plants that are in direct contact with an electrode. When the electrodes touch the plants, the high-voltage DC flows through the leaves and stems into the roots of the plants and finally into the soil. The continuous electrode contact method, and so Electroherb™ technology, is primarily not selective, but with appropriate electrode and applicator designs, selectiveness can be achieved through mechanically insulating the targeted area (e.g. for weed control in row crops), once the application of electrical energy is dependent on the electrodes physically touching the targeted plants.

In terms of an appropriate electrode design of Electroherb™ technology, the possibilities are tremendous, and modular. Besides the electrode–plant contact properties (which optimize electrical conductivity), the type of electrode (e.g. aboveground metal plates or belowground horizontal guiding wheels) determines the amount of electrical energy transferred into plants, as part of the resistive system where the energy applied is consumed. Such limitations, due to high electrical impedance, can be overcome by an increase in voltage. A belowground electrode can guarantee earthing in soil depth with better electrical conductivities, below the dry soil surface. However, such an earthing wheel will not have the same positive effects on vegetation control at

DOI: 10.1079/9781836992288.0005

high plant densities—for instance, two sets of electrodes physically touching the plant as a conductor to provide a "double treatment" of single plants by introducing electrical energy into the plant (positive electrodes; and taking it up from already treated plants). Such a double treatment may reduce the required energy but can only be applied in dense plant populations.

Consequently, a capable, sufficient, and efficient electrical vegetation control device must be adapted to site and crop specific properties to allow the user to reliably apply it for their purposes. On the other hand, the continuous electrode contact method assures that the mode of action, "the electrical agent," is complementary with all electrode settings.

Basic Electrode Types

In electrode design and application practice, metal plate electrodes at low (A) and high (B) plant densities, as well as an alternative electrode type (C) which penetrates the soil, and establishes better earthing properties.

Based on the configurations depicted in (Fig. 5.1) and common electrical weeding setups, here's how the three electrode arrangements can be described.

Plant–soil configuration (positive as active, negative as earthing)

- The positive electrode is applied directly to the plant, delivering an electrical charge that travels through the plant structure.
- The negative electrode (earthing) is connected to the soil, ensuring that the electric current flows from the plant into the ground.
- This setup allows the electrical current to affect both the plant and its root system, as the energy passes through the plant to the soil, where it dissipates.

Plant–plant configuration (positive and negative electrodes on separate plants)

- Both positive and negative electrodes are applied to different plants.
- The current flows between the two electrodes through the plant tissue, ensuring that both plants receive the electrical effect.
- This configuration is typically used to maximize weed-killing efficiency, particularly in areas with dense vegetation, where electricity is passed from one plant to another.

Positive as active with a penetrating negative electrode (disc as earthing)

- The positive electrode is applied to the plant, introducing an electrical charge.
- The negative electrode is a disc that penetrates the soil, ensuring effective grounding.
- This setup improves conductivity by reducing soil resistance, making it particularly useful in dry or high-resistance soils, as the disc ensures deeper electrical penetration.

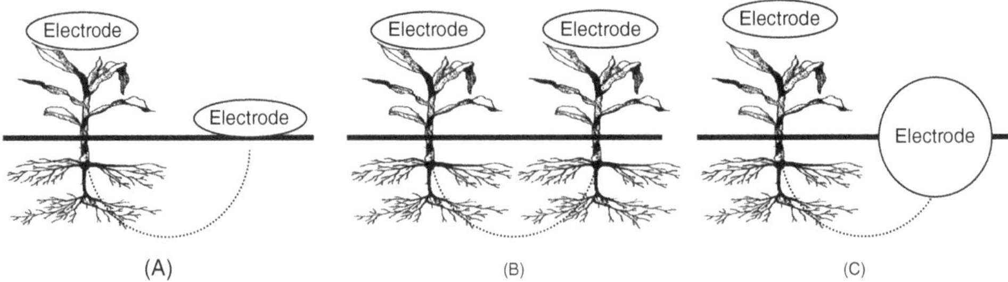

Fig. 5.1. Three main basic electrode types (positive and negative poles can be interchanged): (A) plant–soil configuration (positive as active, negative as earthing); (B) plant–plant configuration (positive and negative electrodes on separate plants); and (C) positive as active with a penetrating negative electrode (disc as earthing).

Each of these configurations has specific applications depending on the weed density, soil conditions, and target efficacy in electrical weeding processes.

Electrode Configurations in Electrical Weeding

Electrode configuration plays a crucial role in the efficacy, efficiency, and safety of electrical weeding applications. The way electrodes are positioned determines how the current flows through the plants and soil, directly impacting weed-killing effectiveness and energy consumption.

Different configurations offer unique advantages and limitations, making them suitable for varied agricultural and environmental conditions. Some setups prioritize energy efficiency by directing most of the current through plants, while others focus on deep-root penetration or safety considerations. The choice of configuration depends on factors such as plant distribution, soil exposure, and the need for precision or broad coverage.

By understanding the strengths and weaknesses of each configuration, users can optimize their application strategy to achieve maximum weed control with minimal energy loss, ensuring a sustainable and effective approach to electrical weeding.

Air–air

The air–air electrode configuration is a highly efficient approach to electrical weeding, designed to optimize energy distribution and maximize plant lethality (Fig. 5.2). Unlike soil-based electrode systems, this method directs the electrical current through the plants twice, ensuring higher energy absorption and minimal loss to the surrounding environment. This results in a more effective weed control strategy, particularly in areas where soil conductivity may otherwise dissipate energy.

However, the efficacy of this system is contingent on the consistent presence of plants between both electrodes (A and B) to complete the electrical circuit, otherwise the maximum voltage would have to be enough to adjust for

Fig. 5.2. Air–air electrode example with average current flow depth.

the implicit higher electrical resistance when secondary plants are absent. Without sufficient vegetation, energy transfer may become less reliable, potentially reducing the overall efficiency of the application.

This method is best suited for environments where vegetation is evenly distributed, such as dense weed patches, uniform crop rows, or areas covered in gravel where soil exposure is minimal. By leveraging the air–air approach, electrical weeding can achieve greater energy efficiency while minimizing unnecessary soil conductivity, making it a practical solution for targeted vegetation control.

Advantages:

- More energy spent on killing plants and less in the soil, since energy goes through the plants twice.

Disadvantages:

- Depends on plant presence at both electrodes (A and B) to ensure good contact.

Best used when there is:

- an even presence of plants throughout the area;
- no soil exposure (e.g. gravel).

Air–soil

The air–soil electrode configuration is a versatile approach to electrical weeding that ensures consistent energy application across different vegetation densities (Fig. 5.3). Unlike the air–air system, this method relies on one electrode making direct contact with the soil, creating a grounded system that enhances stability and safety. The electrical current flows through the

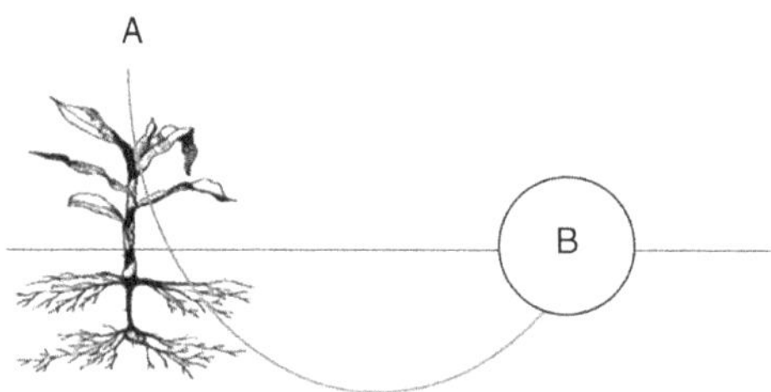

Fig. 5.3. Air–soil electrode example with average current flow depth.

plant once, from the air electrode to the soil electrode, efficiently targeting weeds while minimizing excess energy loss in the ground.

A major advantage of this configuration is its independence from plant distribution—it does not require plants to be present at both electrodes to maintain conductivity. This makes it particularly effective in fields with uneven plant coverage or areas where weeds are sparsely distributed. Additionally, its grounding mechanism enhances safety, ensuring that excess current is dissipated effectively.

Air–soil electrode setup is best suited for agricultural fields with good soil exposure when specific needs such as height selectivity, enhanced depth of current, or stem control are desired.

Advantages:

- Less energy spent on killing plants and less in the soil, since energy goes through the plants once.
- Does not depend on plant presence to ensure good contact.
- Safer, since it is always properly earthed.

Disadvantages:

- Can diminish overall resistance of the system if there are not enough plants, or if the plant distribution is not even.

Best used when there is:

- un-even presence of plants throughout the area;
- good soil exposure, as in regular farming.

Shallow electrode

The shallow electrode system is designed for applications where less soil disturbance is essential (Fig. 5.4). Unlike deeper electrode configurations, the current in a shallow electrode system remains closer to the soil surface, ensuring effective weed elimination while minimizing interference with deeper soil layers.

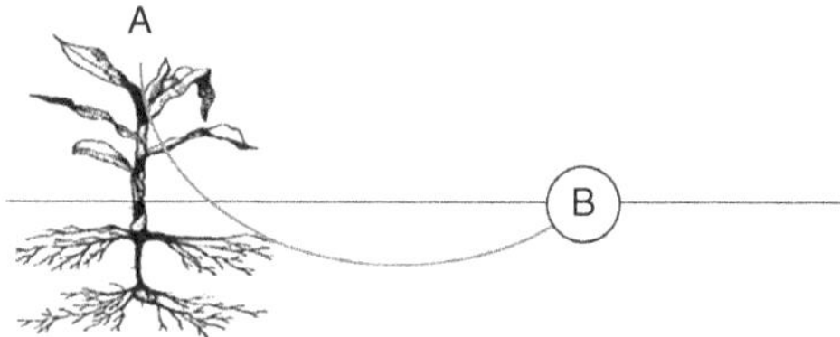

Fig. 5.4. Shallow electrode example with average current flow depth.

One of its key advantages is the reduced structural stress on the applicator, leading to less drag and greater durability over time. Additionally, since it does not penetrate deeply into the soil, it avoids excessive soil disruption, making it ideal for use near main crop roots or in row applicator systems.

However, due to its shallower current penetration, this method may be less effective against weeds with deep root systems. For this reason, it is best utilized in areas without larger plants, where surface-level weed control is sufficient, or in sensitive agricultural settings where soil disturbance must be minimized.

Advantages:

- less soil interference;
- less drag;
- higher durability;
- less structural stress.

Disadvantages:

- Current is shallower and might not reach deeper roots.

Best used when:

- a location is without the presence of larger plants;
- row applicators that go near main crop roots;
- soil disturbance is an issue.

Deep electrode

The deep Electrode system is designed to deliver greater current penetration, effectively targeting deep-rooted weeds and ensuring thorough vegetation control (Fig. 5.5). By directing electricity

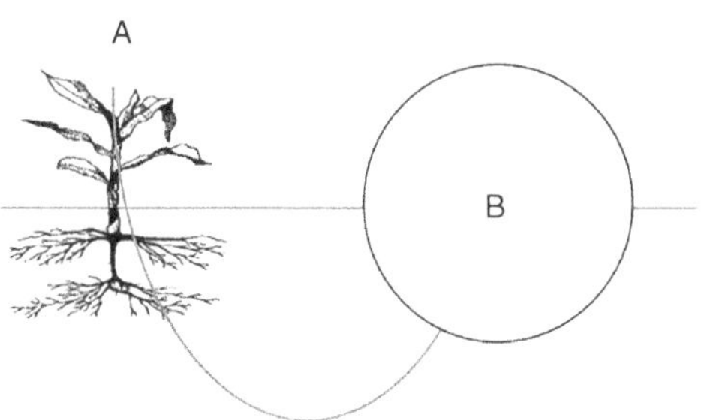

Fig. 5.5. Deep electrode example with average current flow depth.

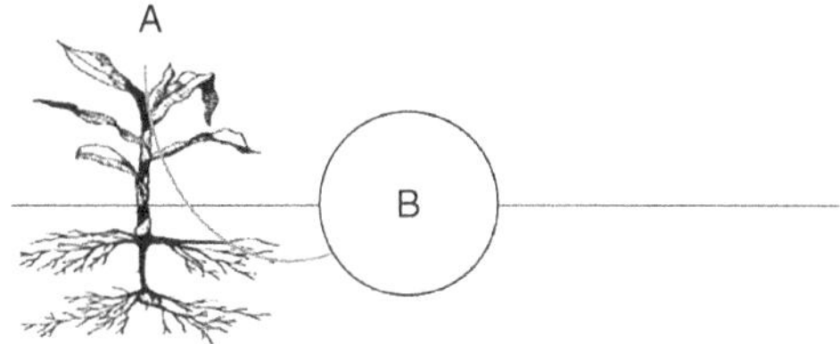

Fig. 5.6. Near electrode example with average current flow depth.

further into the soil, this method increases the likelihood of reaching and neutralizing deeper root systems, making it particularly advantageous for large or more established plants.

This approach is well-suited for applications in open areas, rough terrain, and locations farther from main crops, where soil disturbance is less of a concern. However, deeper electrode penetration introduces more soil interference, increased drag, and greater structural stress on the equipment, which can lead to reduced durability over time.

Advantages:

- Current goes deeper and might reach deeper roots.

Disadvantages:

- More soil interference and drag.
- Diminished durability and more structural stress.

Best used when:

- A location is with the presence of larger plants;
- applying further from main crops or open areas;
- Rough terrain;
- soil disturbance is not an issue.

Near electrode

The near electrode system is highly effective in environments where rapid and efficient weed control is required, making it a strong choice for precision agriculture and controlled applications where soil energy loss must be minimized (Fig. 5.6).

This method is particularly useful in areas without large plants, where shallow-rooted weeds can be effectively neutralized without the need for deep current penetration. Its efficiency makes it an excellent choice for applications where precision and energy conservation are priorities.

However, the closer electrode placement increases the risk of voltaic arcs, which can pose operational challenges. Additionally, due to its shallower current penetration, this system may not be as effective for deep-rooted plants.

Advantages:

- less energy lost in the soil;
- higher efficacy and efficiency at application.

Disadvantages:

- increased voltaic arc risk;
- cannot reach deeper roots.

Best used when:

- a location is without the presence of larger plants;
- efficiency and efficacy are a larger issue.

Far electrode

The far electrode system prioritizes safety and deeper root penetration by reducing the risk of voltaic arcs while allowing current to spread further underground (Fig. 5.7). This makes it well-suited for larger plants that require deeper electrical reach for effective control.

However, energy loss in the soil increases, leading to lower efficiency and efficacy compared to closer electrode configurations. Despite this, it remains useful in applications where deep root treatment is needed, and efficiency is not the primary concern.

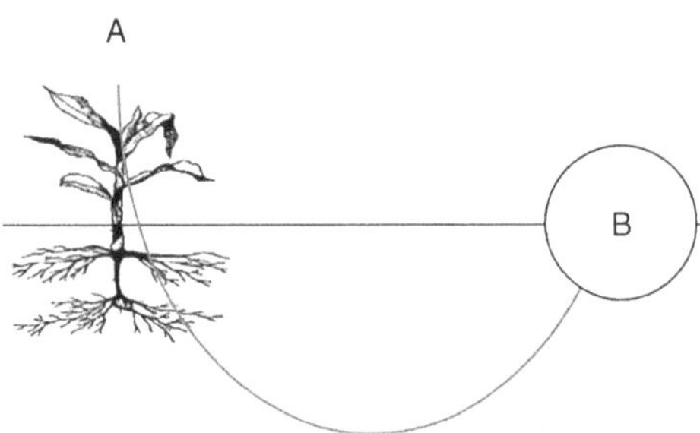

Fig. 5.7. Far electrode example with average current flow depth.

Advantages:

- decreased voltaic arc risk;
- can reach deeper roots.

Disadvantages:

- more energy lost in the soil;
- lower efficacy and efficiency at application.

Best used when:

- a location is with the presence of larger plants;
- efficiency and efficacy are not an issue.

Same plant electrodes

The same plant electrode configuration is designed for targeted desiccation by applying electrical current directly to the aerial parts of the plant, minimizing current flow into the roots or soil (Fig. 5.8). This method ensures rapid drying of the plant, making it particularly effective for preharvest desiccation in crops that require uniform drying before collection.

By focusing energy on the plant's upper structure, this configuration optimizes efficiency and reduces soil energy loss, making it a sustainable alternative to chemical desiccants. However, because it does not impact the root system, it is not suitable for long-term weed control or applications requiring complete plant eradication.

Advantages:

- minimizes current flow into the soil, reducing energy losses;
- targets the plant's aerial system directly, ensuring fast desiccation;
- ideal for preharvest applications where rapid drying is required.

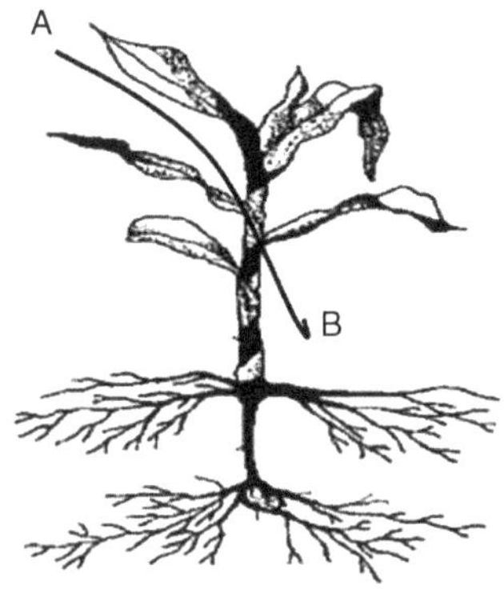

Fig. 5.8. Same plant electrodes example with average current flow.

Disadvantages:

- limited effect on root systems, making it unsuitable for long-term weed control;
- requires precise electrode placement to ensure effective plant contact;
- not effective for deeply rooted perennial weeds.

Best used when:

- preharvest desiccation to dry out plants required before collection;
- root destruction is not necessary;
- crops require controlled drying without soil disturbance.

Electrode-related Application Losses

One of the primary sources of energy loss in electrical weeding applications occurs when an electrode unintentionally makes contact with the soil (Fig. 5.9). This leads to a significant portion of the electrical current bypassing the plants, reducing the overall effectiveness of the treatment.

In the case of electrode A touching the ground, the average current flow depth is altered, allowing energy to dissipate through the soil rather than being fully absorbed by the plant. This not only decreases the weed-killing efficacy but can also result in increased power consumption without proportional weed control benefits. This is a major source of energy waste because, when an "air electrode" touches the ground, it creates an alternative path for the electricity that does

Fig. 5.9. Application losses due to electrode A touching the ground directly with average current flow depth.

not go through the plant itself, therefore being fully wasting the energy. Minimizing these losses requires careful electrode placement and design, ensuring optimal contact with plants while avoiding unnecessary energy dissipation into the soil.

Embodiment Examples

A mechanical embodiment of an electrical weeding device is a physical implementation of the electrical weeding system that is designed to effectively deliver electrical energy to weeds while minimizing harm to crops or other plants.

In a mechanical embodiment of an electrical weeding device, there are typically one or more electrical electrodes that are designed to be placed in close proximity to the weeds, while minimizing contact with the surrounding plants. These electrodes are connected to a power source that delivers an electrical current to the weeds, causing damage to the plant tissue and ultimately killing the weed.

The mechanical embodiment of the electrical weeding device may be handheld, tractor-mounted, or custom-built depending on the specific application. For example, a handheld electrical weeding device may consist of a battery-powered unit that includes a handle, a set of electrodes, and controls for adjusting the electrical energy delivered to the weeds.

A tractor-mounted electrical weeding device may be designed to attach to the tractor and deliver electrical energy to weeds as the tractor moves through the field. The device may include multiple sets of electrodes that can be adjusted to target specific rows or areas of the field.

Custom-built electrical weeding devices may be designed for specific applications, such as vineyard or orchard management. These devices may include specialized electrodes or delivery systems that are optimized for the specific plant type and growing conditions.

Overall, a mechanical embodiment of an electrical weeding device is designed to effectively deliver electrical energy to weeds while minimizing harm to crops or other plants. The design and implementation of the device will depend on the specific application and the needs of the user.

Area applicators

An area applicator for electrical weeding and desiccation of crops, such as soybeans, typically consists of a metal chassis with two or more insulated electrodes that are designed to be placed in close proximity to the crop. In a two-electrode design, the electrodes are typically arranged in a parallel configuration with a gap of several centimeters between them.

To ensure proper insulation and safety, the electrodes are typically made of a conductive material, such as copper or stainless steel, that is coated with an insulating material such as rubber or plastic. The insulation is designed to prevent electrical arcing between the electrodes and other parts of the electrical and electronic equipment, as well as to prevent electrical shock to the user.

Typically, the area applicator includes a power source, such as a battery or generator, that delivers electrical energy to the electrodes. The power source may include safety features, such as overcurrent and overvoltage protection, to prevent damage to the equipment and ensure safe operation.

To ensure effective weeding and desiccation, the area applicator may include a control system that adjusts the amount of electrical energy delivered to the electrodes based on factors such as soil moisture, crop type, and growth stage. The control system may include sensors and feedback mechanisms that allow for real-time adjustment of the electrical energy delivered to the electrodes.

Overall, a well-designed area applicator for electrical weeding and desiccation of crops, such as soybeans, should prioritize safety and effective performance. Insulation of the electrodes from the chassis and other equipment is critical to ensuring safe operation, while a well-designed control system can help to optimize the effectiveness of the weeding and desiccation process.

Area applicators have been developed in three basic kinds, the traditional vertical area applicator, the flexible pulled applicator, and the height selectivity applicator.

Traditional vertical area applicators

Traditional vertical area applicators are versatile tools designed for effective electrical weed control (EWC) across various terrains (Fig. 5.10). Their adaptability allows them to be customized for different applications, making them one of the first widely accepted and utilized designs in the market, especially in organic farming and urban weed management in Brazil.

These applicators consist of three essential components:

- **Electrodes**. The primary element responsible for delivering electrical energy into the plants.
- **Insulating elements**. These components maintain a consistent distance between the electrodes while providing dielectric insulation from the chassis, ensuring safety and efficiency.
- **Solid chassis**. This serves as the structural framework, linking the applicator to its support system.

The primary function of the electrodes is to deliver a high-voltage electrical discharge to the weeds, disrupting their cellular structure and ensuring effective plant desiccation.

Due to their modular design, these applicators can be adjusted in various ways, including horizontal movement, angle modification, and segmentation into individualized units to maintain optimal contact with the ground. These features improve application efficiency and consistency, even in uneven or challenging landscapes. Over time, advancements in insulation materials and electrode configurations have enhanced their durability and performance. Their broad applicability and early market adoption have made them a staple in electrical weeding technology, offering an effective, nonchemical alternative for sustainable vegetation management.

Flexible pulled applicator

The flexible pulled applicator is an advanced weeding tool designed for efficient and uniform electrical weeding, particularly in desiccation applications (Fig. 5.11).

It consists of three main components:

- **Electrodes**. The primary element responsible for delivering electrical energy into the plants.

Fig. 5.10. Traditional vertical area applicators (Zasso).

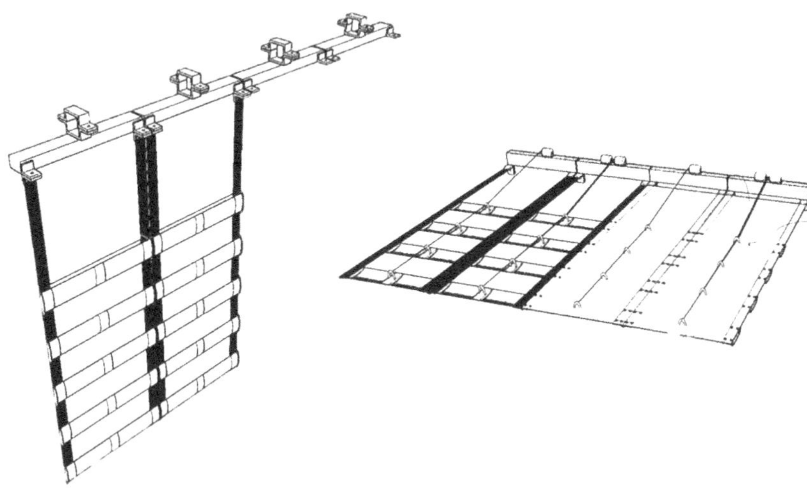

Fig. 5.11. Flexible pulled applicator.

- **Flexible insulating elements.** These components maintain a consistent distance between the electrodes while providing insulation from the chassis, ensuring safety and efficiency.
- **Solid chassis.** This serves as the structural framework, linking the applicator to its support system.

Additionally, the design includes optional features that enhance usability and safety:

- **Retractable pulleys.** These facilitate compact transport by enabling the contraption to be pulled back.
- **Insulating separate cable.** This ensures safe energy transmission through designated cable passings.
- **Insulating protection mats.** These prevent accidental electrical discharge and enhance operator safety.

This applicator is particularly effective in high-density weed coverage, where energy losses to the ground are minimized as most electricity flows directly through the plants. It is also ideal for crop desiccation, such as soybeans, ensuring minimal losses by reducing the risk of dislodging seeds into the soil. The lightweight nature of the electrode further enhances its suitability for delicate desiccation processes.

An embodiment of this design has been described in the patent PCT/BR2021/050249.

Segmented or inter-row applicators

Inter-row applicators are specialized tools designed for targeted weed control between the rows of crops after emergence (Fig. 5.12). These applicators play a crucial role in precision agriculture, offering an effective and sustainable alternative to chemical herbicides.

They consist of a main chassis that allows for easy lifting and transportation, ensuring adaptability to different field conditions. The key functional components of these applicators are the individual segments, which are specifically designed to treat weeds between crop rows. Each segment includes:

- **Insulating careen.** This acts as a dielectric barrier, protecting the main crop from unintended electrical exposure while allowing selective weed targeting.
- **Electrodes.** These components deliver high-voltage energy to invasive weeds, ensuring effective control without disturbing the main crop.
- **Insulating component.** This part isolates the electrodes from the chassis, ensuring safety and proper energy transfer.

Some advanced variations of inter-row applicators offer adjustable electrode width, making them adaptable to different crop-row spacings. Additionally, many designs include pressure-regulating mechanisms to ensure continuous and reliable contact with the soil, which enhances efficiency.

For increased precision, vision-based systems can be integrated to reduce energy loss, especially in fields with low weed density. These systems prevent energy from being wasted on bare soil, directing it exclusively to unwanted vegetation, thereby maximizing efficiency and minimizing unnecessary power consumption.

Vertical stem applicators

The use of vertical electrodes in electrical weeding represents an efficient and precise method for targeted stem-based weed control (Fig. 5.13). These electrodes are particularly effective in postharvest applications, where remaining plant stems can serve as conductors to transfer energy to root-borne organisms, further enhancing weed eradication. Unlike horizontal electrodes, which may allow significant current dissipation through the soil, vertical electrodes focus electrical flow through the plant itself, ensuring a higher energy efficiency. This directed approach also reduces soil disturbance, making it an environmentally friendly alternative for weed management.

Fig. 5.12. Segmented or inter-row applicators.

Fig. 5.13. Vertical stem applicators.

The effectiveness of vertical electrodes stems from their ability to direct electric current into a plant's natural conductive tissues, such as the xylem and phloem. These tissues provide low-impedance pathways, allowing electricity to travel efficiently through the plant's internal structures and maximize vascular system damage, leading to its desiccation and death. Each segment includes:

- **Chassis/frame**. This provides structural support and allows attachment to tractors or machinery.
- **Vertical electrodes**. These deliver high-voltage electricity directly to plant stems for efficient weed control.
- **Insulating support structures**. These prevent electrical leakage and maintain proper electrode positioning.
- **High-voltage power supply**. This generates and regulates the electrical current applied to the electrodes.
- **Contact pressure mechanisms**. These ensure consistent electrode contact with plant stems for optimal energy transfer.
- **Ground wheels/depth control**. Maintains proper electrode height and pressure against the plants.

Spot treatment electrodes

Spot treatment electrodes are designed to selectively apply high-voltage electrical energy to individual plants or small groups. Unlike broad-area applicators, these electrodes focus the electrical discharge on a targeted plant, ensuring maximum efficiency and reducing unintended effects on surrounding vegetation.

Spot treatment electrodes consist of a minimum of two poles: a primary active electrode and a grounding or secondary electrode. The primary electrode directs the high-voltage energy into the target plant, while the secondary electrode provides the return path, ensuring that the electrical current flows through the plant's structure.

Types of spot treatment electrodes

Several electrode configurations have been developed to optimize efficiency and selectivity in different field conditions.

DIRECT CONTACT ELECTRODES. These electrodes must physically touch the plant to create an electrical circuit. Common designs include metal plates, rods, or blade-like structures that ensure a firm connection with the plant's stem or leaves. The high-voltage energy enters through the primary electrode, passes through the plant tissue, and exits via the grounding electrode.

PENETRATIVE ELECTRODES FOR ROOT CONTACT. Some spot treatment electrodes include a perforation system that allows the electrode to penetrate the soil to reach the plant's root zone (Fig. 5.14).

This configuration is particularly useful for deep-rooted weeds, ensuring that the entire plant structure is affected. The electrode's conductive tip is often surrounded by an insulating structure to prevent unintended discharge into the soil.

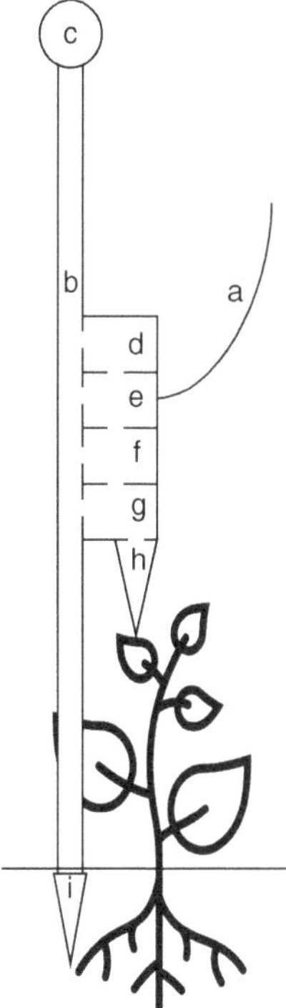

Fig. 5.14. Spot treatment electrode example from a handheld penetrative electrode configuration. (Design from Patent BR 10 2013 007771 2.)

AIR-TO-SOIL ELECTRODES. This method relies on a high-voltage discharge between an aboveground electrode and the soil. The electrical current travels through the plant when it bridges the gap, creating a controlled spark that initiates cellular breakdown. While effective for certain weed species, this method requires precise energy management to avoid excessive arcing.

Safety for spot treatment electrodes

Spot treatment electrodes require rigorous safety measures to prevent unintended electrical hazards. Key safety features include:

- **Insulated housing.** This ensures that only the target plant receives the electrical discharge.
- **Automatic cutoff mechanisms.** These devices shut off when not in contact with a plant to prevent accidental discharge.
- **Operator protection.** Grounding mechanisms and insulated handles prevent accidental shocks.

Drones

Farmers are exploring robotic process automation with image processing, pattern recognition, and machine learning to enhance efficiency in precision agriculture. Drones and satellites enable remote sensing, with drones becoming more capable and cost-effective for complex terrains. Tethered drone assemblies can be vehicle-based or free-standing, with multiple drones tethered by cords to improve environmental surveying (Fig. 5.15). These drones may receive power and communication from a main product tank and operate via sensors or remote control.

A new development replaces traditional dispensing systems with a drone-based electrocution device for invasive plant control. This system consists of a mobile carrier, power electronic converter, and electrodes, with at least one movable electrode directed by a drone. Various embodiments include:

- The drone carrying electrodes while power remains on the carrier, transferring high voltage through cables.
- The drone carrying both electrodes and power electronics.

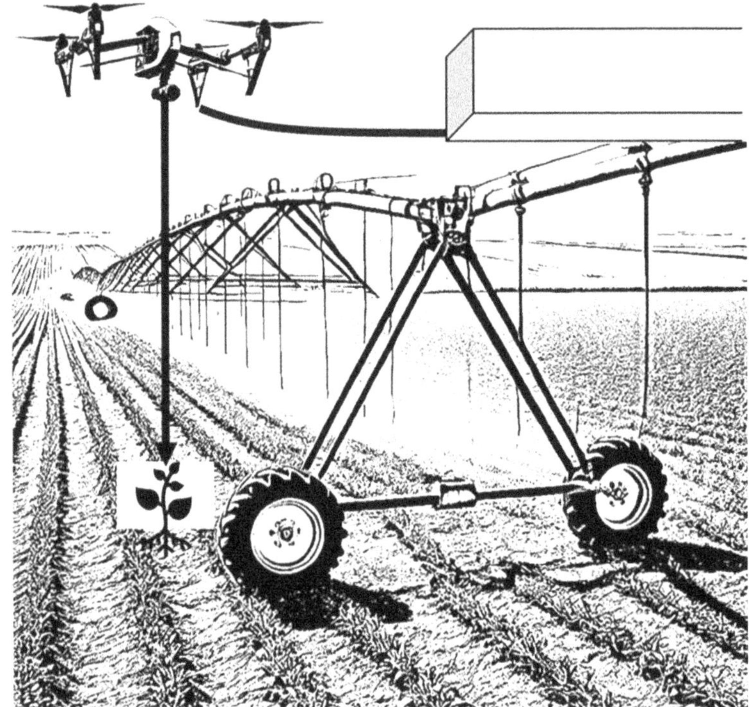

Fig. 5.15. Tethered drone embodiment example.

- Camera sensors mapping precise application areas.
- A tethered drone receiving power through the tether or a parallel cable.
- The drone using platform-based electrical connections or recharging for application.

Optimum Number of Rows of Electrodes

In this analysis, the potential of a third row of electrodes and of earth electrodes is shown using simplified plant–soil models. The complex soil structure, as it occurs in reality, is broken down into easily calculable models. The analysis only includes the parameters under consideration. Simplifications are made to show certain dependencies. These are to be questioned in further investigations. Parameters which are not considered are assumed to be constant: *ceteris paribus*.

The ratio of soil-to-plant resistance is abbreviated with the variable x in order to analyze the comparison of different arrangements as a function of simplified soil properties (Eqn 5.1):

$$x = \frac{R_S}{R_P}$$

$$R_S = x * R_P$$

Equation 5.1. Ratio of soil-to-plant resistance.

The basis for the calculation of the current flow per plant are the characteristics of the constant power sources used. The characteristics can be changed by modular design (parallel connection of several converters). If high-voltage converters are operated in parallel, the resistance range is reduced as the power is constant in relation to the maximum system voltage. For high-power systems, the characteristic is similar to a regulated constant voltage source. The threshold resistance, which defines the transition from constant power to constant voltage source, is calculated from (Eqn 5.2):

$$R_{th} = \frac{(U_{max})^2}{P_{max}}$$

Equation 5.2. Threshold resistance.

First, the simplified equivalent circuit diagrams of different arrangements are shown and how the electrical power is generated in the plant or plants. A distinction is made between arrangements with and without an earth electrode. Earth electrodes are electrodes which ideally do not touch the plant but have direct contact with the soil. The advantage of an earth electrode is the reduction of the load resistance and thus the increase of the current flow. The disadvantage is that this electrode cannot be used to apply a plant.

On the one hand, the calculated maximum peak power which is converted in a plant is described. A plant can be applied several times by passing over the substrate. For arrangements where this is the case, the total power converted is also described. If two electrodes are used, for example, the total power is twice the peak power, because the plant is applied twice with the same current. Temporal changes of the electrical conductivities are initially neglected here.

Two electrodes

The arrangements with two electrodes are shown below.

With earth electrode

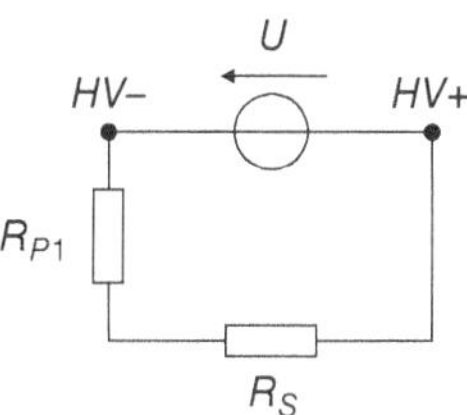

Schematic 5.1. Two-electrode arrangement with earth electrode schematic.

Constant voltage source (Eqn 5.3):

$$P_{RP1} = I_{RP1}^2 * R_{P1} = \left(\frac{U_{const}}{R_{P1} + R_S}\right)^2 * R_{P1}$$
$$= \left(\frac{U_{const}}{R_P * (1+x)}\right)^2 * R_P$$

Equation 5.3. Two-electrode arrangement with earth electrode Schematic 5.1 with a constant voltage source.

Constant power source (Eqn 5.4):

$$P_{const} = P_{RP1} + P_{RS} = P_{RP1} + I_{RP1}^2 * R_S =$$
$$P_{RP1} + \frac{P_{RP1}}{R_{P1}} * R_S = P_{RP1} + R_{RP1} * x = P_{RP1} *$$
$$(1+x)\ P_{RP1} = \frac{P_{const}}{1+x}$$

Equation 5.4. Two-electrode arrangement with earth electrode schematic with a constant power source.

Without earth electrode

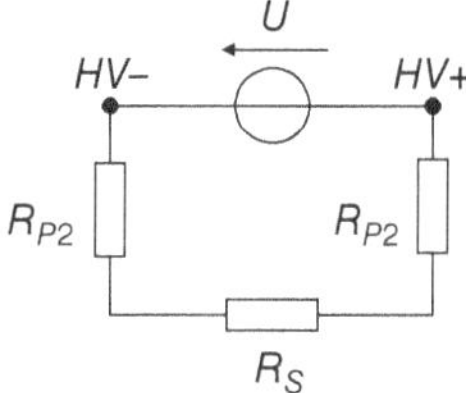

Schematic 5.2. Two-electrode arrangement without earth electrode schematic.

Peak power converted in a plant

Constant voltage source (Eqn 5.5):

$$P_{RP2} = I_{RP2}^2 * R_{P2} = \left(\frac{U_{const}}{2 * R_{P2} + R_S}\right)^2 * R_{P2}$$
$$= \left(\frac{U}{R_P * (2+x)}\right)^2 * R_P$$

Equation 5.5. Peak power per plant with constant voltage source.

Constant power source (Eqn 5.6):

$$P_{const} = 2 * P_{RP2} + P_{RS} = P_{RP2} * (2+x)$$
$$P_{RP2} = \frac{P_{const}}{2+x}$$

Equation 5.6. Peak power per plant with constant power source.

Total power converted in a plant

Constant voltage source (Eqn 5.7):

$$P_{RP2} = 2 * I_{RP2}^2 * R_{P2} = 2 * \left(\frac{U_{const}}{2 * R_{P2} + R_S}\right)^2 * R_{P2}$$
$$= 2 * \left(\frac{U_{const}}{R_P * (2+x)}\right)^2 * R_P$$

Equation 5.7. Total power converted into a plant with constant voltage source.

Constant power source (Eqn 5.8):

$$P_{RP2} = 2 * \frac{P_{const}}{2+x}$$

Equation 5.8. Total power converted into a plant with constant power source.

Three electrodes

The arrangements with three electrodes are shown below.

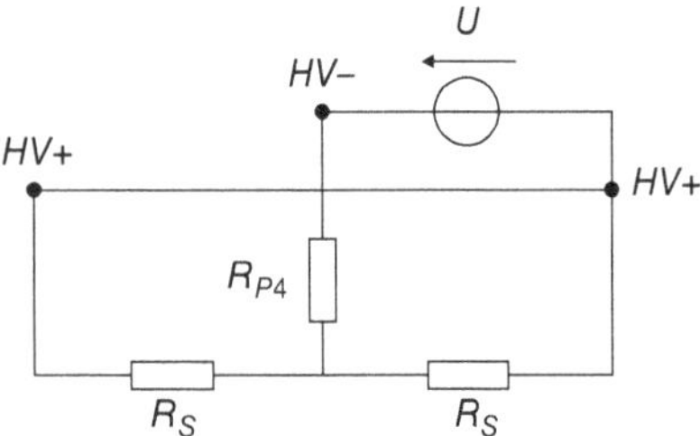

Schematic 5.3. Three electrodes with earthing.

With earth electrodes

Constant voltage source (Eqn 5.9):

$$P_{RP4} = I_{RP4}^2 * R_{P4} = \left(\frac{U_{const}}{R_{P4} + 0{,}5 * R_S} \right)^2 * R_{P4}$$

$$= \left(\frac{U_{const}}{R_P * (1 + 0{,}5 * x)} \right)^2 * R_P$$

Equation 5.9. Total power converted into a plant with three-electrode applicators and a constant voltage source.

Constant power source (Eqn 5.10):

$$P_{const} = P_{RP4} + 2 * P_{RS} = P_{RP4} + 2 * I_S^2 * R_S = P_{RP4} + 2 * \left(\frac{I_{RP4}}{2} \right)^2 * R_S = P_{RP4} + \frac{1}{2} * I_{RP4}^2 * x * R_P$$

$$= P_{RP4} + \frac{1}{2} * P_{RP4} * x = P_{RP4} * \left(1 + \frac{1}{2} * x \right)$$

$$\text{therefore} \quad P_{RP4} = \frac{P_{const}}{1 + \frac{1}{2} * x}$$

Equation 5.10. Total power converted into a plant with three-electrode applicators and a constant power source.

Without earth electrodes

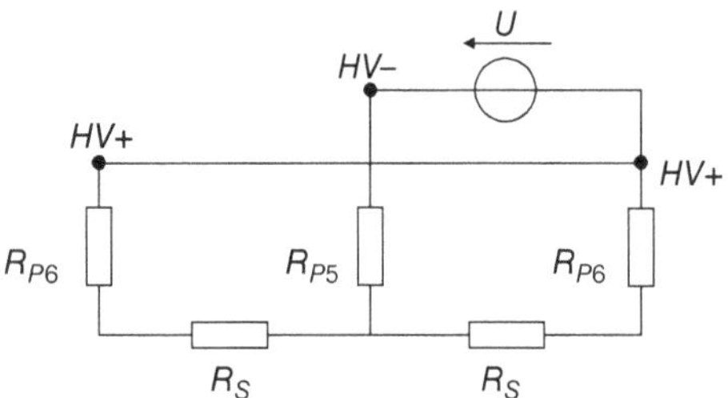

Schematic 5.4. Three-electrode schematics without earthing.

Peak power converted in a plant

Constant voltage source (Eqn 5.11):

$$P_{RP5} = I_{RP5}^2 * R_{P4} = \left(\frac{U_{const}}{R_{P5} + 0{,}5 * (R_S + R_{P6})} \right)^2 * R_{P5}$$

$$= \left(\frac{U_{const}}{R_P * (1{,}5 + 0{,}5 * x)} \right)^2 * R_P$$

$$P_{RP6} = I_{RP6}^2 * R_{P6} = \left(\frac{1}{2} * I_{RP5} \right)^2 * R_{P6}$$

$$= \frac{1}{4} * \left(\frac{U_{const}}{(R_{P5} + 0{,}5 * (R_S + R_{P6}))} \right)^2 * R_{P6}$$

$$= \frac{1}{4} * \left(\frac{U_{const}}{R_P * (1{,}5 + 0{,}\ 5 * x)} \right)^2 * R_P$$

Equation 5.11. Peak power converted into a plant with three-electrode applicators without earthing and a constant voltage source.

Constant power source (Eqn 5.12):

$$P_{const} = P_{RP5} + 2 * (P_{RS} + P_{RP6})$$

$$= P_{RP5} + 2 * (I_{RS}^2 * R_S + I_{RS}^2 * R_P)$$

$$= P_{RP5} + 2 * \left(\left(\frac{I_{RP5}}{2} \right)^2 * x * R_P + \left(\frac{I_{RP5}}{2} \right)^2 * R_P \right)$$

$$= P_{RP5} + \frac{1}{2} * (P_{RP5} * x + P_{RP5})$$

$$= P_{RP5} * \left(1{,}5 + \frac{1}{2} * x \right)$$

$$P_{RP5} = \frac{P_{const}}{1{,}5 + \frac{1}{2} * x}$$

$$P_{RP6} = I_{RP6}^2 * R_{P6} = \left(\frac{I_{RP5}}{2} \right)^2 * R_P = \frac{1}{4} * P_{RP5}$$

Equation 5.12. Peak power converted into a plant with three-electrode applicators without earthing and a constant power source.

Total power converted in a plant

Constant voltage source (Eqn 5.13):

$$P_{tot} = P_{RP5} + 2 * P_{RP6} = \frac{3}{2} * \left(\frac{U_{const}}{R_{P5} + 0,5 * (R_S + R_{P6})} \right)^2 * R_{P5} = \frac{3}{2} * \left(\frac{U_{const}}{R_P * (1,5 + 0,5 * x)} \right)^2 * R_P$$

Equation 5.13. Total power converted into a plant with three-electrode applicators without earthing and a constant voltage source.

Constant power source (Eqn 5.14):

$$P_{tot} = P_{RP5} + 2 * P_{PR6} = P_{RP5} + 2 * \frac{1}{4} * P_{RP5} = \frac{3}{2} * P_{RP5} = \frac{3}{2} * \frac{P_{const}}{1,5 + \frac{1}{2} * x}$$

Equation 5.14. Total power converted into a plant with three-electrode applicators without earthing and a constant power source.

Mathematical Comparison

The following analysis was done between the author and Zasso's electrical engineering team, including the good mathematical work of Christopher Freinmann. Different arrangements are compared qualitatively. Since both the peak power and the total power may have an influence on the quality of the application, they are compared separately. Both the influence of one or more earth electrode(s) and the influence of a third row of electrodes are considered. Furthermore, a distinction is made between constant voltage source and constant power source.

In other words, we will now explore the mathematical comparison of different features of potential applicators. The graph itself is the relationship of power of these different features, with the objective of setting the best possible configuration given the different possibilities. The power relationships are stated in the Y axis, and the resistance relationships in the X axis, and the comparisons are made both for peak and total power consumed by the target plants.

The feature changes regard the presence of an earthing electrode, and the presence of two or three electrodes. This study is key to understanding when to use each type of electrode configuration, and the advantage that may be expected by using each one. On top of that, it also is made visible how the relationship of the resistances of plants and soil may affect the power at the plant level.

As limitations, this mathematical study does not consider the differences between air system and root system resistances at each individual plant level, unwanted losses such as active electrodes touching the ground, or current leaving the plant throughout its root system, and not only at the end.

Also, it does not consider that these systems are imperfect: plants touch each other; one electrode touches more than one plant at once; plants are not fully in series with the soil; as current flow from a plant throughout its root systems, the root systems are not a continuous resistance, but branch out in different ways for different species; mowed plants may have root system parts that are not connected to the air parts; different morphologies affect how the system works; there are no perfect "voltage" and "power" sources, as in real case scenarios it is common that they present a "power" or "voltage" curve regarding load; etc.

So, these are mathematical studies of specific different variables that can be controlled at the applicator level, to be validated in the field, by practical research. Therefore, it must be considered that this is (and will always be) an imperfect mathematical study that collapses many variables that are, in the field of realities, ranges of possibilities, into simplified values; and does not consider many relevant features that will likely vary from practical case to case.

That said, at constant (or at least nearly comparable) scenarios regarding soil resistance, plant morphology, power and voltage sources, and other relevant variables, the result of this study should have a positive correlation between the studied scenarios and expected averages, at least in stable and comparable scenarios. Constant voltage systems benefit more significantly from additional electrodes and earth electrodes, while constant power systems provide more consistent energy delivery, reducing the need for additional electrodes while still maintaining good efficiency. In summary, constant power systems are more resilient to configuration changes, making them the better choice for optimizing electrical weeding performance across different conditions.

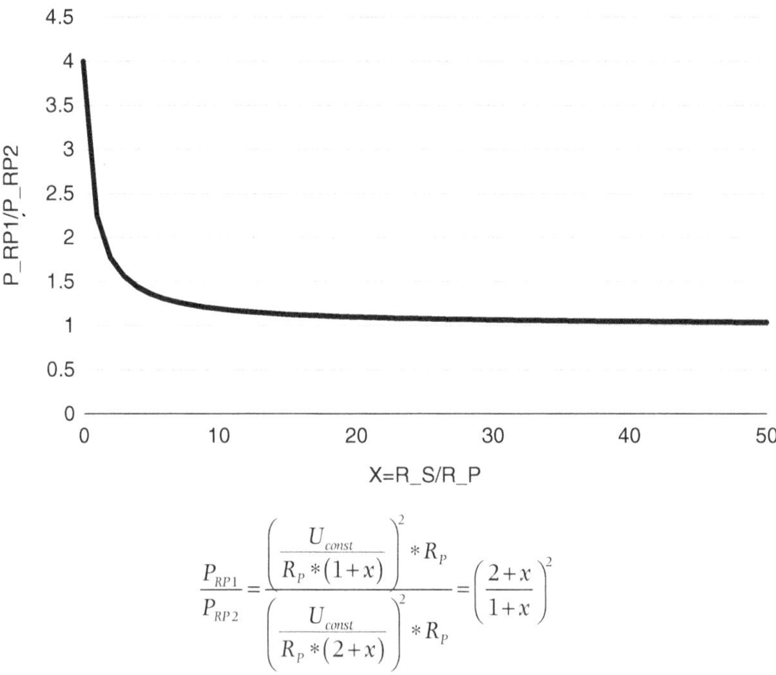

$$\frac{P_{RP1}}{P_{RP2}} = \frac{\left(\frac{U_{const}}{R_P * (1+x)}\right)^2 * R_P}{\left(\frac{U_{const}}{R_P * (2+x)}\right)^2 * R_P} = \left(\frac{2+x}{1+x}\right)^2$$

Equation 5.15. Peak power comparison: constant voltage source. Advantage of grounding electrode reducing load resistance.

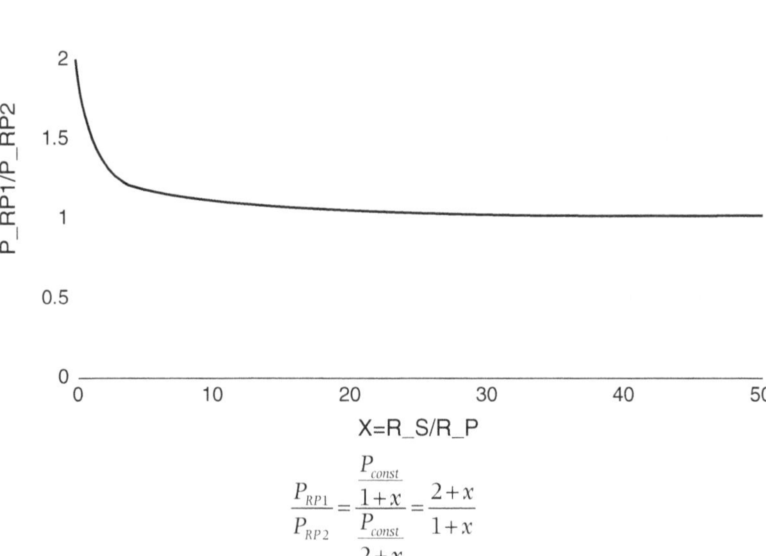

$$\frac{P_{RP1}}{P_{RP2}} = \frac{\frac{P_{const}}{1+x}}{\frac{P_{const}}{2+x}} = \frac{2+x}{1+x}$$

Equation 5.16. Peak power comparison: constant power source. Advantage of grounding electrode reducing load resistance.

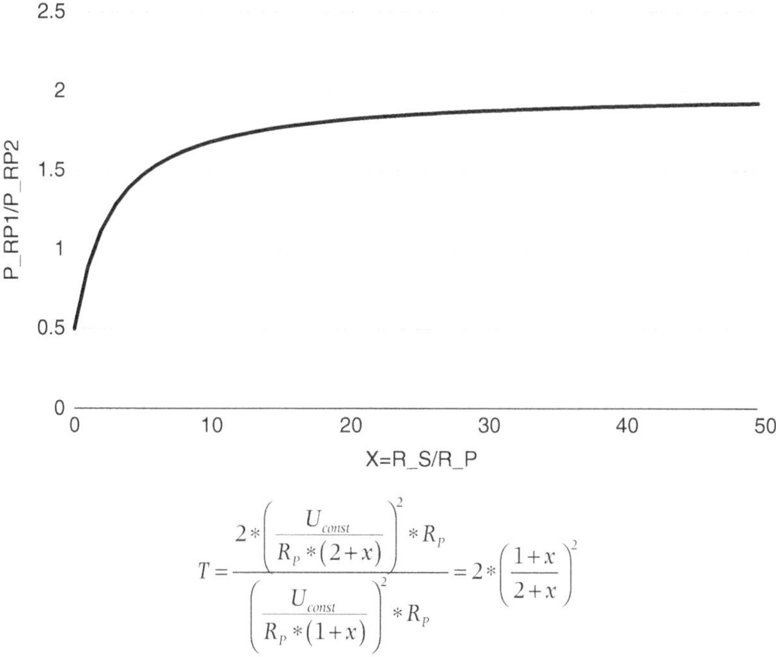

$$T = \frac{2 * \left(\frac{U_{const}}{R_p * (2+x)} \right)^2 * R_p}{\left(\frac{U_{const}}{R_p * (1+x)} \right)^2 * R_p} = 2 * \left(\frac{1+x}{2+x} \right)^2$$

Equation 5.17. Total power comparison: constant voltage source. Advantage of two applying electrodes.

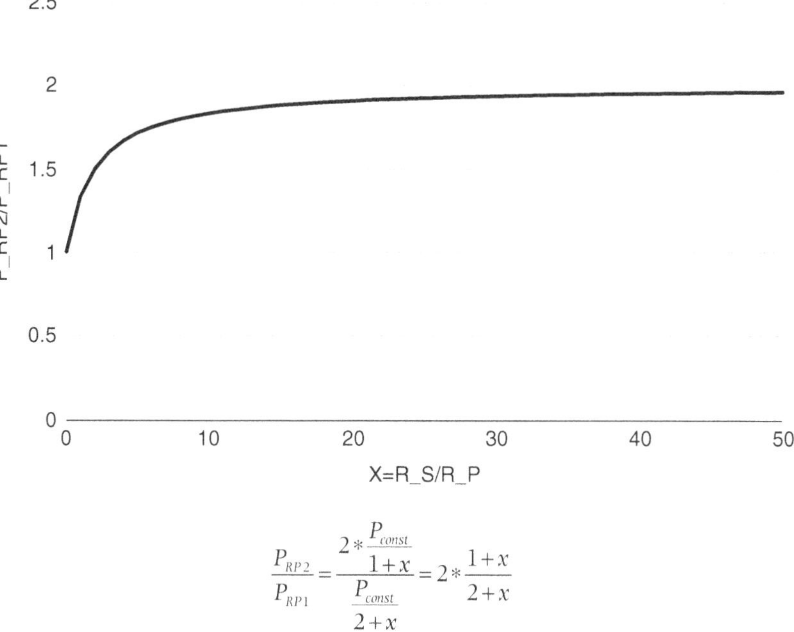

$$\frac{P_{RP2}}{P_{RP1}} = \frac{2 * \frac{P_{const}}{1+x}}{\frac{P_{const}}{2+x}} = 2 * \frac{1+x}{2+x}$$

Equation 5.18. Total power comparison: constant power source. Advantage of two applying electrodes.

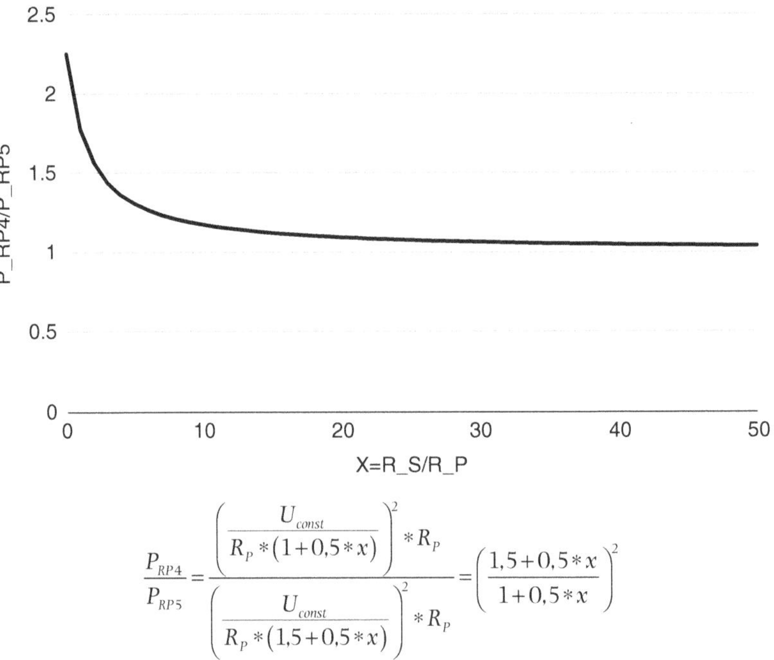

$$\frac{P_{RP4}}{P_{RP5}} = \frac{\left(\frac{U_{const}}{R_p*(1+0{,}5*x)}\right)^2 * R_P}{\left(\frac{U_{const}}{R_p*(1{,}5+0{,}5*x)}\right)^2 * R_P} = \left(\frac{1{,}5+0{,}5*x}{1+0{,}5*x}\right)^2$$

Equation 5.19. Peak power comparison: constant voltage source. Advantage of the earth electrodes.

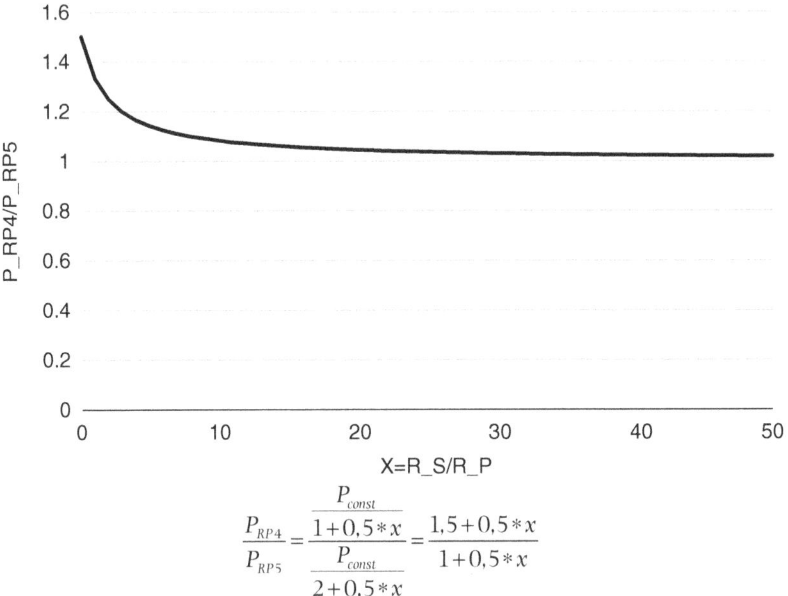

$$\frac{P_{RP4}}{P_{RP5}} = \frac{\frac{P_{const}}{1+0{,}5*x}}{\frac{P_{const}}{2+0{,}5*x}} = \frac{1{,}5+0{,}5*x}{1+0{,}5*x}$$

Equation 5.20. Peak power comparison: constant power source. Advantage of the earth electrodes.

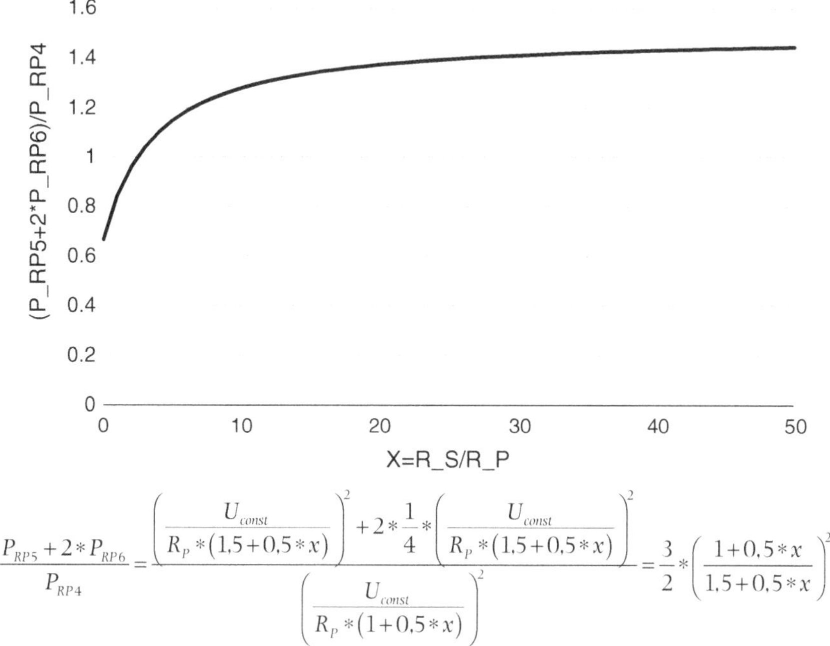

$$\frac{P_{RP5}+2*P_{RP6}}{P_{RP4}}=\frac{\left(\frac{U_{const}}{R_P*(1{,}5+0{,}5*x)}\right)^2+2*\frac{1}{4}*\left(\frac{U_{const}}{R_P*(1{,}5+0{,}5*x)}\right)^2}{\left(\frac{U_{const}}{R_P*(1+0{,}5*x)}\right)^2}=\frac{3}{2}*\left(\frac{1+0{,}5*x}{1{,}5+0{,}5*x}\right)^2$$

Equation 5.21. Total power comparison: constant voltage source. Advantage of three touches with three applying electrodes.

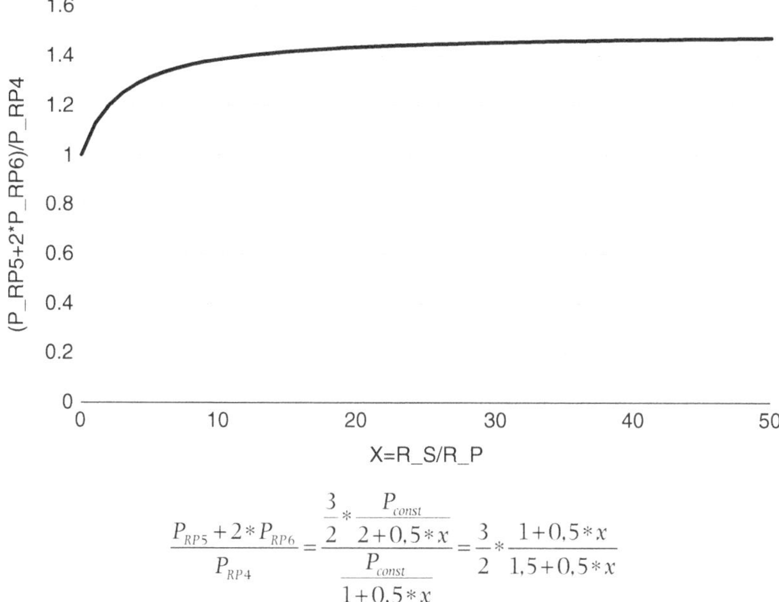

$$\frac{P_{RP5}+2*P_{RP6}}{P_{RP4}}=\frac{\frac{3}{2}*\frac{P_{const}}{2+0{,}5*x}}{\frac{P_{const}}{1+0{,}5*x}}=\frac{3}{2}*\frac{1+0{,}5*x}{1{,}5+0{,}5*x}$$

Equation 5.22. Total power comparison: constant power source. Advantage of three touches with three applying electrodes.

Two vs three electrodes: With earth electrodes

Advantage of a third electrode (a second earth electrode).

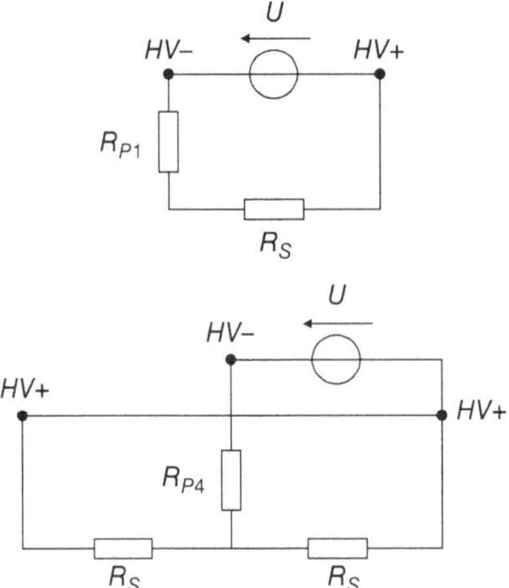

Schematic 5.5. Two vs three electrodes: with earth electrodes.

Constant voltage source (Eqn 5.23):

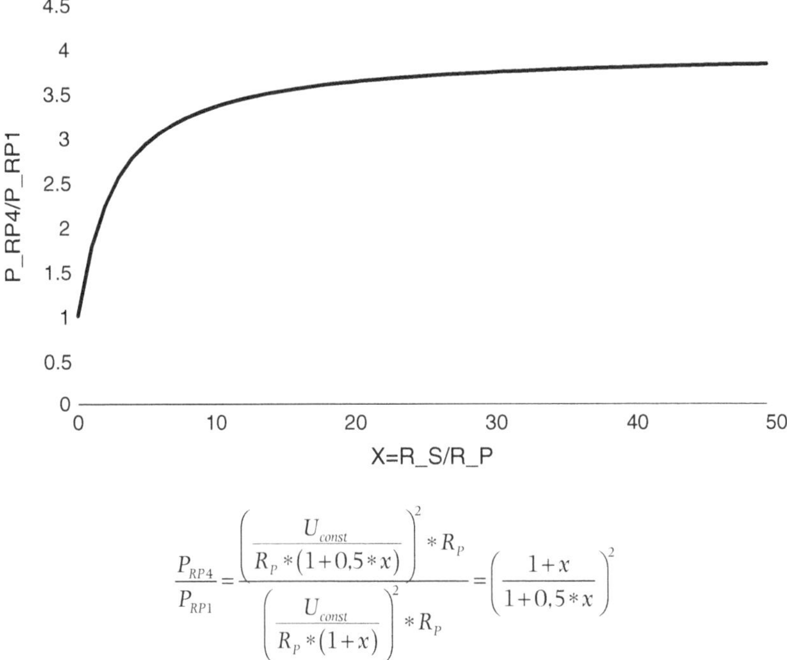

$$\frac{P_{RP4}}{P_{RP1}} = \frac{\left(\frac{U_{const}}{R_p * (1+0{,}5*x)}\right)^2 * R_p}{\left(\frac{U_{const}}{R_p * (1+x)}\right)^2 * R_p} = \left(\frac{1+x}{1+0{,}5*x}\right)^2$$

Equation 5.23. Power advantage of a third electrode (a second earth electrode) at a constant voltage source.

Constant power source (Eqn 5.24):

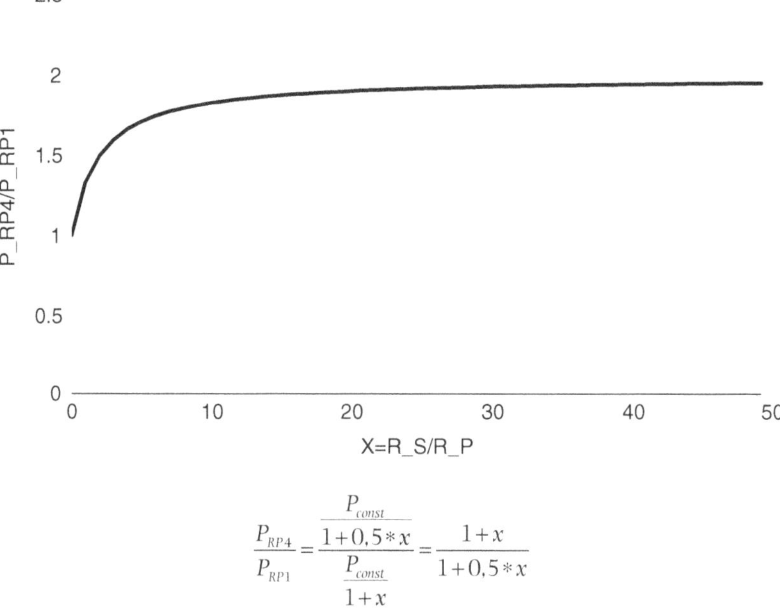

$$\frac{P_{RP4}}{P_{RP1}} = \frac{\dfrac{P_{const}}{1+0,5*x}}{\dfrac{P_{const}}{1+x}} = \frac{1+x}{1+0,5*x}$$

Equation 5.24. Power advantage of a third electrode (a second earth electrode) at a constant power source.

Two vs three electrodes: Without earth electrodes

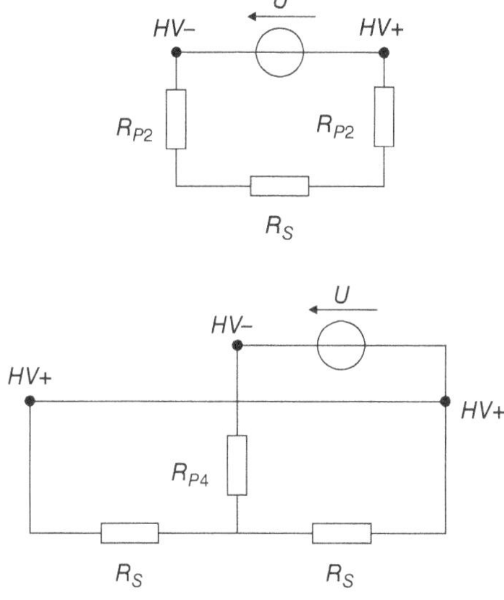

Schematic 5.6. Two vs three electrodes: without earth electrodes.

Constant voltage source (Eqn 5.25):

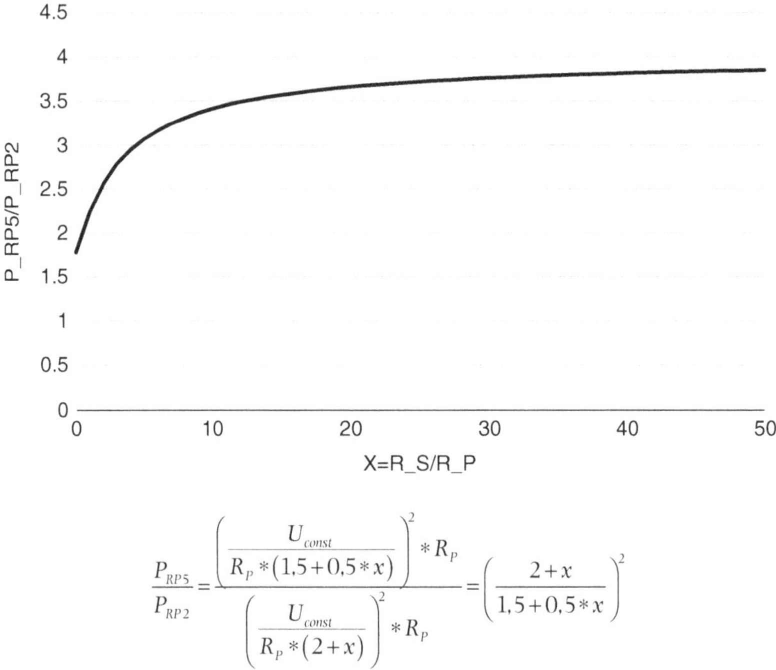

$$\frac{P_{RP5}}{P_{RP2}}=\frac{\left(\frac{U_{const}}{R_P*(1{,}5+0{,}5*x)}\right)^2*R_P}{\left(\frac{U_{const}}{R_P*(2+x)}\right)^2*R_P}=\left(\frac{2+x}{1{,}5+0{,}5*x}\right)^2$$

Equation 5.25. Peak power comparison: advantage of a third electrode at a constant voltage source.

Constant power source (Eqn 5.26):

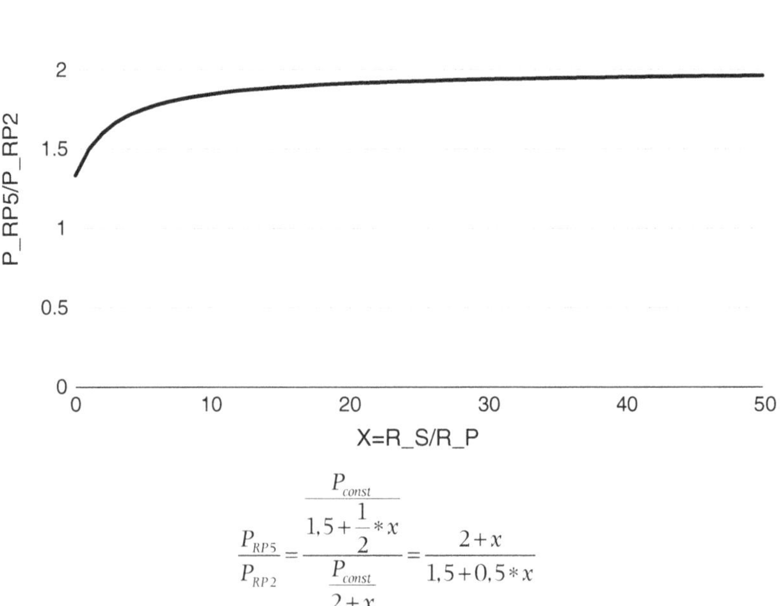

$$\frac{P_{RP5}}{P_{RP2}}=\frac{\frac{P_{const}}{1{,}5+\frac{1}{2}*x}}{\frac{P_{const}}{2+x}}=\frac{2+x}{1{,}5+0{,}5*x}$$

Equation 5.26. Peak power comparison: advantage of a third electrode at a constant power source.

Constant voltage source (Eqn 5.27):

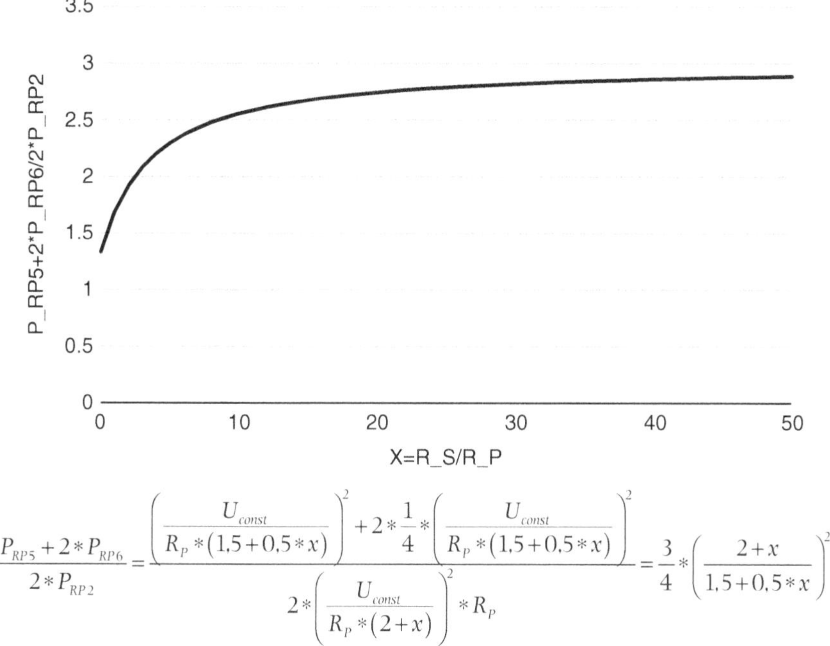

$$\frac{P_{RP5}+2*P_{RP6}}{2*P_{RP2}}=\frac{\left(\frac{U_{const}}{R_P*(1{,}5+0{,}5*x)}\right)^2+2*\frac{1}{4}*\left(\frac{U_{const}}{R_P*(1{,}5+0{,}5*x)}\right)^2}{2*\left(\frac{U_{const}}{R_P*(2+x)}\right)^2*R_P}=\frac{3}{4}*\left(\frac{2+x}{1{,}5+0{,}5*x}\right)^2$$

Equation 5.27. Total power comparison: advantage of a third electrode at a constant voltage source.

Constant power source (Eqn 5.28):

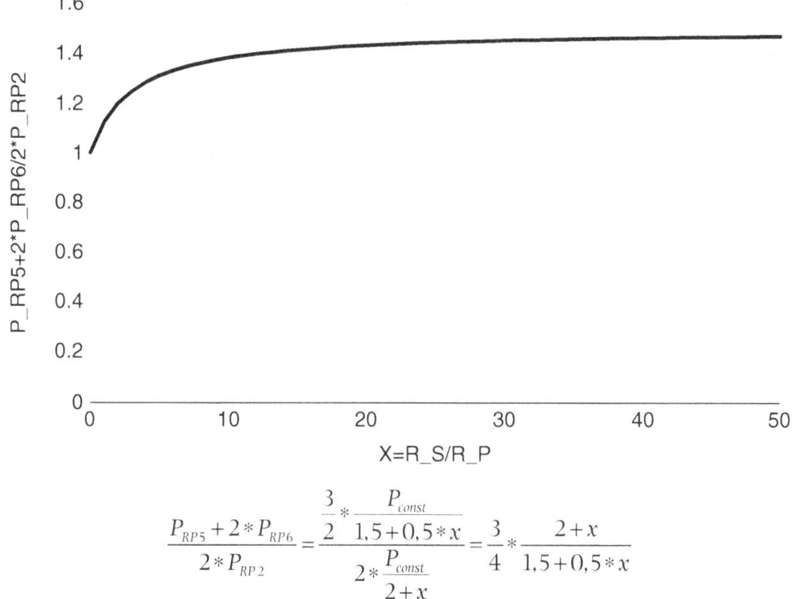

$$\frac{P_{RP5}+2*P_{RP6}}{2*P_{RP2}}=\frac{\frac{3}{2}*\frac{P_{const}}{1{,}5+0{,}5*x}}{2*\frac{P_{const}}{2+x}}=\frac{3}{4}*\frac{2+x}{1{,}5+0{,}5*x}$$

Equation 5.28. Total power comparison: advantage of a third electrode at a constant power source.

Conclusions for a constant voltage source

Table 5.1 compares different electrode configurations under a *constant voltage source*, analyzing their impact on *peak power* and *converted total power*.

The key conclusions that can be drawn from this comparison are:

- **Earth electrodes improve efficiency but have limits.**
  - When using *two electrodes*, adding an *earth electrode* can improve performance *up to a factor of 4* in terms of converted total power.
  - The improvement is slightly better in peak power but is not as dramatic as in total power.
  - However, in the case of *three electrodes*, the earth electrode provides a *maximum improvement of up to 2.25 times* in peak power and *1.5 times in converted power*, which is less significant than in the two-electrode case.
- **Three electrodes perform better than two without earth electrodes.**
  - When comparing *two vs three electrodes without an earth electrode*, three electrodes perform better in both peak and converted total power.
  - The improvement factor for peak power is at least *1.77*, while for converted total power, three electrodes perform *up to 3 times better*.
- **Three electrodes show strong advantages in many scenarios.**
  - In general, *three electrodes outperform two electrodes* in most conditions.
  - The performance gain ranges from *a factor of 1.33 to 4*, depending on the specific power condition and whether an earth electrode is used.

Table 5.1. Constant voltage source comparison summary.

Constant voltage source													
Two electrodes: with vs without earth electrode				Three electrodes: with vs without earth electrode				Two vs three electrodes: without earth electrode				Two vs three electrodes: with earth electrodes	
Peak Power		Converted total power		Peak power		Converted total power		Peak Power		Converted total power			
X<10	X>10	X<10	X>10	X<10	X>10	X<10	X>10	X<10	X>10	X<10	X>10	X<10	X>10
Ground electrode up to factor 4 better	Earth electrode slightly better	Earth electrode slight advantages	Without earth electrode up to factor 2 better	Earth electrodes up to factor 2.25 better	Earth electrode slightly better	Earth electrode slight advantages	Without earth electrode at least factor 1.5 better	Without earth electrode at least factor 1.77 better	Three electrodes up to factor 4 better	Three electrodes at least factor 1.33 better	Three electrodes up to factor 3 better	Advantage three electrodes, only for X = 0 equally good	Three electrodes up to factor 4 better

- **Earth electrodes offer small to significant gains depending on configuration.**
    - In some cases, adding an earth electrode provides only *slight advantages*, particularly when using *three electrodes*.
    - In other cases (especially with two electrodes), an earth electrode can significantly boost performance *up to a factor of 4*.
- **For two vs three electrodes with earth electrodes, differences are situation-dependent.**
    - *At X = 0 (specific condition not detailed in the table), two and three electrodes perform equally well.*
    - Otherwise, three electrodes *can improve performance by a factor of up to 4*, making them a more robust choice under most conditions.

Conclusions for a constant power source

Table 5.2 compares different electrode configurations under a *constant power source*, analyzing their impact on *peak power* and *converted total power*.

The key conclusions that can be drawn from this comparison are:

- **Earth electrodes provide moderate advantages.**
    - When using two electrodes, an earth electrode *improves performance by up to a factor of 2* in converted total power.
    - The earth electrode gives only slight improvements in peak power *but is still beneficial.*
    - In the case of three electrodes, earth electrodes *provide up to 1.5 times better*

Table 5.2. Constant power source comparison summary.

Constant power source													
Two electrodes: with vs without earth electrode				Three electrodes: with vs without earth electrode				Two vs three electrodes: without earth electrode				Two vs three electrodes: with earth electrodes	
Peak Power		Converted total power		Peak power		Converted total power		Peak Power		Converted total power			
X<10	X>10	X<10	X>10	X<10	X>10	X<10	X>10	X<10	X>10	X<10	X>10	X<10	X>10
Ground electrode up to factor 2 better	Earth electrode slightly better	Advantage without earth electrode; for X = 0 equally good	Without earth electrode up to factor 2 better	Earth electrodes up to factor 1.5 better	Earth electrode slightly better	Advantage without earth electrode; for X = 0 equally good	Without earth electrode at least factor 1.5 better	Without earth electrode at least factor 1.33 better	Three electrodes up to factor 2 better	Advantage three electrodes; for X = 0 equally good	Three electrodes up to factor 1.5 better	Advantage three electrodes, only for X = 0 equally good	Three electrodes up to factor 2 better

performance in converted total power but do not significantly improve peak power.
- **Three electrodes improve performance compared to two electrodes.**
  - Without earth electrodes, three electrodes can *perform up to 2 times better* than two electrodes.
  - With earth electrodes, three electrodes still provide *a factor of 1.5 improvement* in converted total power over two electrodes.
  - However, in specific conditions (x = 0), three electrodes perform equally well as two electrodes, meaning that their *advantage depends on specific power settings.*
- **Earth electrodes have limited impact compared to constant voltage systems.**
  - Compared to the constant voltage system, where earth electrodes provided a greater improvement (up to a factor of 4), in a constant power system, the improvement is much lower (maximum factor of 2).
  - This suggests that under a constant power system, the ability to adjust power delivery mitigates some of the advantages of additional earth electrodes.
- **Three electrodes vs two electrodes: earth electrodes are less influential.**
  - When comparing two vs three electrodes with earth electrodes, the three-electrode system *improves performance by a factor of 2* in converted total power.
  - However, for specific conditions (x = 0), two and three electrodes perform equally well.

Spraying Liquids to Lower Resistance

Since 2011, Zasso (then operating as Sayyou) has actively explored the potential of conductive liquid spraying as a method to improve the performance of EWC (Fig. 5.16). The first field tests were conducted in sugarcane fields in Goiasa, Goiatuba, Brazil, with the goal of evaluating whether spraying could enhance the effectiveness of electrical weed destruction.

Original hypothesis and expected benefits

The theoretical basis for integrating conductive liquid spraying into the electrical weeding process was rooted in three main potential benefits:

- **Reduction of electrical resistance.** By introducing a conductive medium, the assumption was that electrical resistance across the plant system could be lowered, allowing higher energy penetration into the plant and roots, increasing overall weed control efficiency.
- **Improved energy transfer and system efficiency.** With lower resistance, it was expected that less energy loss would occur, optimizing power usage and reducing the overall energy demand required for plant destruction.
- **Fire prevention.** One of the persistent challenges in electrical weeding, particularly in dry vegetation, is the potential for fires caused by excessive localized heating. Spraying a controlled conductive liquid was theorized to mitigate the risk of ignition, ensuring a safer operational environment.

Practical results and unexpected outcomes

Despite promising initial assumptions, real-world field tests revealed a series of drawbacks that ultimately made conductive liquid spraying an impractical solution for Zasso's electrical weeding technology.

Limited impact on system efficiency and efficacy

- While conductive liquids did lower electrical resistance in some cases, the improvement was not significant enough to justify its use.
- When properly designed electrodes were used, the gain from spraying became negligible, making it an unnecessary addition.
- The additional operational cost of liquid preparation, storage, and application outweighed any potential performance gains.

Fig. 5.16. Spraying in combination with electrical weeding (Sayyou, 2011).

Inferior fire control compared to alternative systems

- Although conductive liquid spraying helped reduce fire risks in some conditions, it was not a reliable method for fire prevention.
- Other fire control mechanisms, including voltage and power regulation, electrode optimization, and insulation improvements, proved to be far more effective and significantly cheaper.
- The superiority of alternative solutions was later formalized in patent EP 20 788 713.4, which presents more efficient, cost-effective, and scalable fire prevention strategies.

Severe safety risks for operators and equipment

- One of the most critical issues encountered was the increased human and equipment safety risk.
- In an electrical weeding system, applying a conductive liquid introduces the risk of creating unintended electrical paths, particularly

if the liquid's dielectric breakdown strength is lower than the applied voltage.

- *Potential hazards include*:
  - **Electrical discharges to bystanders or operators.** If the liquid inadvertently creates a conductive bridge between the electrical system and a person, it could result in severe electric shock or even fatal consequences.
  - **Equipment damage and short circuits.** The sprayed liquid could accidentally form conductive paths inside the machinery, leading to short circuits, malfunctions, or costly damage to sensitive electronic components.
- Although safety measures could theoretically be implemented to mitigate these risks, they would introduce unnecessary complexity and further increase operational costs, making the approach impractical.

Conclusion: A discontinued concept due to practical challenges

Although conductive liquid spraying was a promising concept in early theoretical models, field experience demonstrated that it was not a viable solution due to:

- Limited benefits in efficiency and efficacy compared to proper electrode design.
- Higher operational costs without proportional performance gains.
- Inferior fire prevention performance compared to other methods.
- Significant safety risks for both operators and equipment, making the system difficult to implement safely at scale.

While not entirely impossible to use conductive spraying in a controlled manner, its drawbacks outweigh its benefits, and Zasso has since focused on more effective and safer alternatives for optimizing EWC.

Spraying Liquids: Effects on Resistance, Efficiency, and Efficacy

The thesis disproved in the previous failed Zasso (Sayyou at the time) 2011 try-outs with liquid spraying was that nondeionized water, and other similarly conductive liquids, would increase electrical weeding efficiency and efficacy through reduced electrical resistance. Unfortunately, this became a failed experiment for several reasons.

Filling tanks is a costly operation. A spraying operation takes between 100–1,000 l, and usually about 250 l ha^{-1}. For an average farm size of about 180 ha in the United States (and about 50 ha in Brazil), this would mean an average of about 45–50 refilling operations with a 1,000 l tank. Every refilling operation means lost time, lost water, and lost fuel through crossing the farm to the refilling point.

As an example, for relatively low-electrical-power equipment that can weed at a rate of 1 ha h^{-1} with a 1,000 l tank, which serves about 4 ha, and considering that filling operations take 30–60 min to complete, 25% losses on time efficiency would be incurred, not including further water and fuel losses. Of course, this can be minimized with more powerful equipment with larger tanks, but, under no circumstances, is this scalable enough to become negligible, as the weight and volume of water carried scales proportionally further, increasing fuel costs and the need for hardware investments.

As shown before, the theoretical electrical efficiency of an electrical system is related to the resistances between electrodes and the resistive system of plants and soil. Although water or other conductive liquids may, indeed, reduce the electrical resistance between the electrode and the target plants, practical experiments dating from 2011 have shown that it actually decreases energy efficiency and the efficacy of application.

The results of the failed experiments are in line with several physical facts:

- Spraying creates a non-negligible mass that is heated during the operation, creating heat losses and, therefore, loss of energy efficiency.
- If water distribution is not optimized, it can conduct energy through unwanted paths that deviate the energy from the targeted plant directly to the soil, and this effect is enhanced if high frequency AC is used, due to electrical skin effect, where most energy flows to the external part of the conductor—in this case, the liquid coating created by spraying.

- Diminished impedance does not increase electrical energy applied when using newer constant power converters instead of older constant voltage converters that have lower operational efficacy.
- Any gain on lowering impedance can be matched by proper electrode design that increases the contact area and/or time between the applicators and targeted plant.

This has been tried and improved by Zasso for many years, since 2011 (when it was still called Sayyou). Unfortunately, the unavoidable conclusion is that, with a proper electrode design that matches the field plant type, density, and soil resistance, there is an efficiency and efficacy loss on using a costly resort that increases equipment and human safety concerns for electrical weeding.

Spraying Liquids: Effects on Fire Control

Spraying liquid has been tried for sugarcane in Brazil by Zasso since 2011. Although it works, other preferred options, without its intrinsic downsides, are available. A few examples:

- The patent EP 20 788 713.4 proposes the use of an insulating mat that prevents oxygen being feed to dry organic matter in electrical weeding.
- Rounded shapes avoid electromagnetic pressure in the extremities (sparks are usually generated at the "tips"; this is why spark generation tesla coils commonly have a protruding conductive electrode to control where the sparks will fly from).
- Larger electrode contact areas and more electrodes diminish the impedance and, therefore, the applied voltage at constant power converters, reducing the risks of spark formation.
- Pulse width modulation (PWM) of the electrical power source can manage the formation, size, and time of eventual sparks to the time of application continuity with minimal losses, considering the deionization time of air varies between 1–10 min.
- The patent PCT/IB2017/001456 also shows the possibility of using a variety of sensors, such as corona (ultraviolet and others) sensors, to identify the formation of sparks before they occur and using the electronic controls of the electrical converters feeding the applicators to avoid spark formation before or at their formation.
- Older patents also preconize the use of the carbon dioxide (CO_2) generated from the equipment or tractor exhaust to extinguish eventual fire formation, although this was never successfully implemented in commercial use.

Therefore, the conclusion is that although using sprayed liquids to control sparks can be used successfully, there are other preferred methods. While they do work as a fire-risk control, other means to perform control are just as, if not more, efficient, and do not present any efficiency or efficacy losses, increased costs, nor increased safety concerns.

Alternating Current vs Direct Current in Electrical Weeding

Efficacy on systemic plant control is a direct consequence of killing the plant's root system, not the air system. A plant might recover if its air system is somewhat damaged, but it will certainly die if its root system becomes unable to send nutrients to the rest of the plant, that necessarily will perish from nutrient deficiency. Ideally, the efficacy of plant electrocution can be considered to be binary (plant is dead or not dead). Efficacy can be considered as the relation between the fatal energy consumption needed at the individual plant's root and the volume of the root system. This is extensively explained in literature and patents dating back to 1990s and early 2000s (i.e. Brazilian PI 0502291).

The more recent literature (i.e. patent PE3437) relates to the comparison between AC and DC systems in terms of efficacy and efficiency of EWC. It shows that, considering that an electrical weeding system is always constrained by the maximum power output of an AC or DC power source, the ideal system outputs constant power regardless of the total resistance the electrodes experience.

In order to maximize the effectiveness of the treatment, it is necessary to maximize energy consumed, and the energy consumed is the integral of the power curve in time, which is by itself dependent on the voltage. Therefore, the ideal system would always provide DC with a voltage output that ensures constant power, given the ever-changing load of real field operation.

The DC voltage values should preferably be in the range of 1–10 kV, which is the range necessary to ensure constant power delivery in the most common load range of a regular field operation. The peak DC voltage is defined as the maximum DC voltage. In particular cases, such as railway dead beds, highways, and urban roads, these values can reach up to between 15–40 kV.

This means, ideally, that the processed electrical energy output should not be a waveform with repeating shapes, but a DC power source that is able to vary its voltage according to the load, so as to ensure constant power delivery. Given the cycles of charge and discharge of the capacitors of the voltage multiplier, when the load is low, parasite voltage peaks may be seen, but should be kept as low as possible and always below 1 kV.

On an AC system, there are three further issues that cause it to be not the preferred option. The first issue relates to apparent and true power. When DC is applied, the resulting applied power simply follows Ohm's Law, with a theoretically perfect efficiency (resistive heaters are 100% efficient because they convert all electricity into heat). On the other hand, when AC is applied, you may have the effective true power a lot lower than the apparent power. Apparent power is the product of the root mean square (RMS) voltage and the RMS current in an AC circuit. When the impedance is purely resistance, it is equal to true power. However, when reactance is present, apparent power is greater than true power. The power triangle illustrates the relationship between apparent power, true power, and reactive power, which is the vector difference between the two. Reactive power, measured in volt–ampere reactive (VAR), is energy that is stored and then released as a magnetic or electrostatic field. The equation $Pa^2 = Pt^2 + Pr^2$ represents the relationship between apparent power, true power, and reactive power in a complex AC circuit.

The second issue relates to peak-to-peak voltages. Dielectric strength relates to maximum peak-to-peak voltages. RMS voltage of an AC waveform is the amount of AC power that produces the same heating effect as DC power. The RMS voltage of a sinusoidal source of electromotive force (V_{RMS}) is used to characterize the source. It is the square root of the time average of the voltage squared. The value of V_{RMS} is V0/ Square root of$\sqrt{2}$, or, equivalently, 0.707V0.

In other words, as RMS voltage is always lower than peak-to-peak voltage, DC always needs less voltage than AC for the same power delivered and energy consumption. Therefore, there are fewer risks of undesired sparks and lower dielectric strength of insulators needed in the building of the equipment, providing a cheaper and more efficient solution.

The third issue relates to skin effects. This issue applies only to high-frequency AC converters. Skin effect is a phenomenon where the distribution of AC within a conductor becomes concentrated near its surface, resulting in a decrease in current density with increasing depth. This causes the majority of the current to flow at the "skin" of the conductor, between the outer surface and a level called the skin depth. The skin depth is inversely proportional to the frequency of the AC, meaning that as frequency increases, current flow moves closer to the surface, resulting in a smaller skin depth. This effect increases the effective resistance of the conductor by reducing its effective cross-section. It is caused by opposing eddy currents created by the changing magnetic field from the AC. At 60 Hz in copper, the skin depth is around 8.5 mm, and at high frequencies, the skin depth becomes significantly smaller.

This means that as the frequency of the AC increases, less energy will be consumed by the target internal part of the plants where the xylem and phloem are located. This loss is much higher if there are any conductive materials on the outside of the plant, such as moist soil or conductive sprayed liquids.

Therefore, the clear conclusion is that, *ceteris paribus*, a DC is preferred to an AC system for electrical weeding converters.

Electric Weeding vs Steam Weeding

Among the various weed control methods, two environmentally friendly alternatives stand out: electric weeding and steam weeding. Both approaches aim to eliminate weeds without relying on chemical herbicides, but their energy efficiency and overall effectiveness differ significantly.

Using real-world data from operational equipment, the analysis demonstrates that electric weeding is significantly more efficient, consuming 8–16 times less energy per unit area than steam-based methods. This advantage arises from key factors such as speed, energy conversion efficiency, and the fundamental physics of each technique.

One of the most significant differences between electric and steam weeding is the *work rate* (Table 5.3)—the area that can be treated per unit of time.

Electric weeding covers five times the area per hour compared to steam weeding. This means that even if both methods had the same energy requirements per square meter, electric weeding would still be five times faster, leading to more efficient operations and lower labor costs.

Energy efficiency in weeding methods depends on how the input energy is converted into effective weed destruction. Electric weeding uses a high-voltage system to apply an electric discharge directly to the plant. The electrical current *disrupts cell structures*, causing immediate and irreversible damage. The process is highly targeted, with minimal energy lost to the surrounding environment.

Energy per square meter:

- 43 kJ m^{-2} (calculated based on electrical equipment power);
- 69 kJ m^{-2} (accounting for losses from diesel-powered generation).

Steam weeding relies on thermal shock, requiring high-energy steam production. Natural gas is burned to heat water, creating pressurized steam, which is then applied to weeds. However, significant energy is lost in the conversion process, including:

- **Combustion inefficiency.** Only a portion of the energy in natural gas is effectively converted into usable heat.
- **Heat loss during steam transport.** Steam loses heat as it travels through hoses.
- **Energy transfer inefficiency.** Steam must heat plant cells through convective transfer, which is less efficient than direct electric energy application.

Energy per square meter:

- 565 kJ m^{-2} (expected efficiency);
- 678 kJ m^{-2} (assuming 100% efficiency, which is unrealistic).

Table 5.3. Electric weeding vs steam weeding energy consumption.

Method	Work rate	Energy consumption
Electric	5,000 m^{-2}/h @ 1 m width	60,000 W/h
Steam	1,000 m^{-2}/h @ 1 m width	89.3 m^3 natural gas per m^3 water

Table 5.4. Electric vs steam: fuel consumption.

Method	Diesel/Gasoline consumption	Natural gas consumption
Electric	8 liters of diesel per hour	43,400,000 J/l (total energy input)
Steam	0.02 m^{-3} natural gas per hour	31,650,000 J/m^3 natural gas

Table 5.5. Electric vs steam: kJ m^{-2}

Electric		Steam	
43	kJ m^{-2}*	565	kJ m^{-2} @ exp. efficiency
69	kJ m^{-2}**	678	kJ m^{-2}@ 100% efficiency

*Calculated per electrical weeding equipment power.

**Calculated per diesel running the equipment, losses accounted for.

Thus, electric weeding requires at least 8 times less energy than steam weeding at comparable efficiency levels (Tables 5.4 and 5.5).

Despite the lower absolute fuel volume for steam, the conversion inefficiencies result in a higher energy demand per square meter. Steam weeding requires significant amounts of gas per cubic meter of water, further increasing operational costs.

Therefore, according to calculations, it should take between 8–16 times more energy to perform weeding using steam, when compared to electrical weeding.

References

Diprose, M.F. and Benson, F.A. (1984) Electrical methods of killing plants. *Journal of Agricultural Engineering Research* 30(3): 197–209.

Dykes, W.G. (1979) *Plant Destruction Using Electricity. U.S. Patent 4,177,603*. Lasco, Inc., Vicksburg, MS, USA.

Savchuk, V.N. and Bayev, V.I. (1975) An investigation of electrodes for electric spark treatment of plants. In: *Electrochemistry in Industrial Processing & Biology*. Scientific Information Consultants, London, UK, pp. 66–70.

Wilson, R.G. and Anderson, F.N. (1981) Control of three weed species in sugarbeets (*Beta vulgaris*) with an electrical discharge system. *Weed Science* 29(1): 93–98.

6

Underlying Technological Generations

Introduction

What do we mean by Generations (Gens.) 0, 1, 2, and 3?

The purpose of the power module is to deliver electrical energy to the target plants' vascular (electrically resistive) system. The system faces great variability of electrical impedance, due to the different electrode shapes, humidity, soil composition, compactness and weed pressure of the substrate. Therefore, to make the most of the available power, it is necessary to ensure stability of power delivery. For this purpose, different strategies can be implemented, and those strategies can be subdivided into the following broad technological generations:

- **Gen. 0.** Low-frequency (no inverters) voltage supply with manual controls for the voltage (such as variacs or a transformer tap). For academic purposes only, it cannot be transformed into a commercial product due to high operational risk, limited application, and low efficiency.
- **Gen. 1.** Gen. 0 with added features to limit current to avoid catastrophic failures (such as current-limiting reactors [CLRs]), current and continuity controls (such as solid-state relays and similar solutions), and rectifying bridges to minimize the number of electrodes and enhance energy efficiency through direct current (DC) application.
- **Gen. 2.** Higher internal frequency of operation is achieved through inverters that allow for a plethora of different active electronic controls to avoid failure, stabilize power delivered, and maximize operational efficiency. Can be alternating current (AC) or DC.
- **Gen 3.** Gen. 2 with a power electronic architecture that passively ensures no failures, stable power, and operational efficiency without any active control, ensuring minimal cost, volume, and weight. The dynamic self-adjusting system optimizes the application parameters electronically without the need to actively change the system's impedance. Further electronic controls can be implemented if needed to further the precision but are not a necessity.

Generation 1

Usually, Gen. 1 systems were comprised of some or all the following components (Schematic 6.1):

- A sinusoidal power source for the system. This could usually be:
  - a generator group;
  - an alternator coupled to a tractor;

DOI: 10.1079/9781836992288.0006

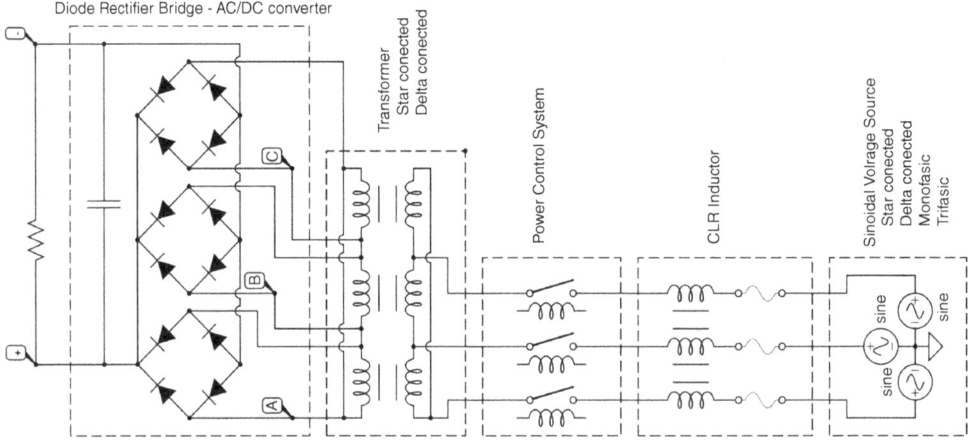

Schematic 6.1. Generation 1 simplified.

- a power outlet;
- others.

- A CLR inductor to protect against overcurrent flows and increase cos φ. This step of the circuit can also contain/include:
  - fuses;
  - circuit breaker;
  - thermomagnetic circuit breaker.
- A power control system comprised of solid-state relays or other pulse width modulation (PWM) or power control components, usually controlled by software or manual calibration.
- Transformer system to ensure proper voltage for electrical weeding—usual power/voltage sinusoidal power sources are in the 100–1,000 V range and electrical weeding requires 1,000–20,000 V.
- Rectification bridge composed of diodes and capacitors to reduce the need for different applicators, enhance weeding efficiency and performance, and make it simpler to mechanically build applicators where the poles are at the same distance (and therefore approximately the same electrical resistance) from each other.

Alternatively, the alternator, CLR, and rectifier can be substituted by one single DC power source or similar.

There is a huge difference in topology between each item of equipment, as can be seen by comparing the electronic circuit shown in Schematic 6.1 and the traditional circuit Schematic 6.2.

An evolution of Gen.1 was presented through the development of the electrode comutation idea (Schwager and Schwager, 2005a, 2005b). Inherently, this basic electrical weeding circuit allows for the potential solution of the issues with previous electrical weeding circuit designs, as previously stated in this section and described in the following sections on Generation 2 and Generation 3.

Generation 2

Because of the electrical (instead of electronic) characteristics of most of the previous systems that worked at relatively low frequencies (usually 50–65 Hz), those systems had very high weight, were very large, and, although robust, did not usually self-adjust to avoid problems, enhance efficiency, and efficacy. Neither could they take advantage of recent developments in new sensors and high-power electronic components, which work at much larger frequencies. Some of the most common issues are as follows:

- The CLR (also fuses and thermomagnetic circuit breakers) provide a strategy to limit current in the system overall, but do nothing to control it, nor to ensure the system works at or near optimum performance and energy consumption.
- Due to the large size of the transformers, it is difficult to control the energy at smaller segments, which would have to be connected to only one transformer each. This causes the application quality to be uneven, especially when some plants provide a low

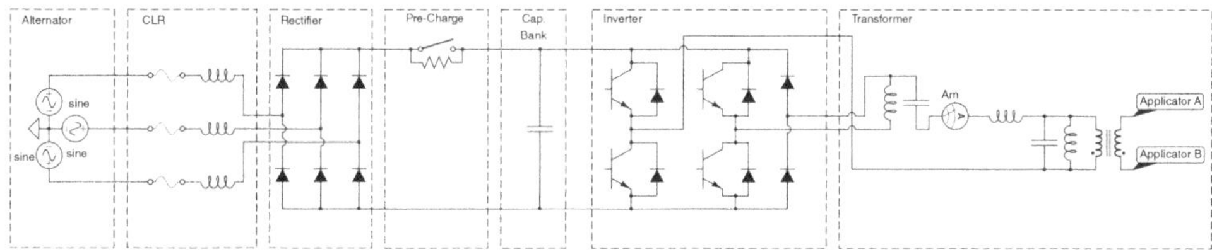

Schematic 6.2. Generation 2 simplified.

electric resistance path, leaving other plants without enough energy for a good quality control across the whole width of the applicator.

- High imperfections in application (some plants were not electrified because of the large "off" time) are inherent in the low frequency of the power control systems, which cannot work in frequencies much higher than the power generation component (50–65 Hz).
- Difficulty adjusting proper current trough PWM, since the low frequency is translated as a high variation of energy consumption at the power generating component (alternator, power outlet, etc.).
- Large cycle and response time, due to low frequency of the system (at the PWM control, which cannot have much higher frequencies than the power source), causes inevitable deviations from the optimum power (and therefore energy, efficiency, and efficacy of the system)—the electrical resistance of the plant–soil system can change more rapidly than the system can self-adjust, especially if such equipment runs faster and/or has a large width.
- Adjustments of the PWM in the frequency range of voltaic-arc (sparks) formation, and greater than the time to deionize the air (0.1–100 ms), can influence the creation of undesired sparks that can start fires if the application is done over large amounts of dry organic matter.
- Due to the inherent size, cost, weight, and nature of the traditional 50–65 Hz transformers, it is an engineering challenge to subdivide the applicator into large numbers of transformers to be controlled individually—this causes uneven applications throughout the width of the system since one part of the applicator can have access to a completely different electrical resistance to another, causing some parts to apply too much power while others apply too little (possibly not enough to kill undesired invasive plants).
- Traditional 50–65 Hz transformers are very heavy and expensive.
- Because of lack of power control and lack of power consumption attenuation, the power consumption can change rapidly and back-influence the frequency of the power generation (frequency of the alternator)—this can not only damage the power generation components, but it can also cause rapid changes in frequency and voltage throughout the system, since some alternators cannot self-adjust for voltage fast enough to account for the change.
- The potential variations in voltage can cause problems or damage, especially at the power control systems (solid-state relays) and rectification bridges.

These problems have different solutions, but it is important to consider that most of the issues arise from the fact that the previous technology was electric, not electronic, without any kind of telemetry or self-regulation.

The hallmark of Gen. 2 technology can be considered to be the patents from 2013–2018 regarding electrical weeding filed by Sayyou-Zasso, such as the extensive PCT/IB2017/001456 (Coutinho Filho *et al.*, 2017).

Frequency influence in efficacy

In the first Sayyou Japanese patent, Mr. Satoru Narita studied plant electrocution up to ~1 kHz and found out that the efficiency increases with frequency up to about the same efficiency of the DC. In recent trials, Sayyou found that it stabilizes at about that same efficiency, even if higher frequencies are used (Sayyou Co., Ltd., 1995).

Empirically, what has been observed is that plants respond in a stable manner in a sigmoidal way after around 0.5–1 kHz, according to the adjusted graph in Fig. 6.1.

For other reasons, DC is preferred to AC in the construction of electrical weeding equipment due to the skin effect, lower peak-to-peak voltages, and apparent vs true power, as explained in the later section in Chapter 6 "Alternating current vs direct current: The power factor."

Electronic weeding circuit

The basic circuit for an electric weeding circuit is comprised of at least two of the following components (Schematic 6.2):

- An alternator or other AC power source.
- A CLR to limit current.
- A rectifier or rectification bridge to provide DC for the DC/AC converter (usually a square-wave h-bridge inverter).
- A capacitor bank to provide and attenuate the DC/AC converter (usually a square-wave h-bridge inverter) energy consumption peaks.
- A transformer.

Solution to high imperfections in application due to the low frequency of control

In previous systems operating in lower frequencies, the current control was traditionally done through the use of solid-state relays. These relays could not operate in frequencies much higher than the AC power source if they were to work inside each cycle (50–65 Hz). This could lead to high imperfections on application (some plants were not properly controlled because of the large "off" time).

As an example: a onetime gap in a 50 Hz cycle in equipment running at 5 km h^{-1} means an off area of 2.77 cm. Considering small plants can are less than 2.77 cm, such a gap would entirely prevent some plants from being controlled through electrocution.

As the current technological stage DC/AC converter (usually a square wave h bridge inverter) allows for frequencies in the order of 1 kHz to 1 MHz, problems related to low frequency of the current control system are not relevant when using such technologies, by using the current control technology proposed here. The idea is to control the PWM at the DC/AC converter, and therefore the overall current of the system, which has never been used before for electrical weeding (Fig. 6.2).

Solution to high variations of energy consumption at the power generating component

Long on-off periods (especially if greater than to 10 ms) can cause a huge difference in power consumption at the power generating component. This difference can be translated into mechanical and electrical component stress and loss due to heat. Long off periods to control for

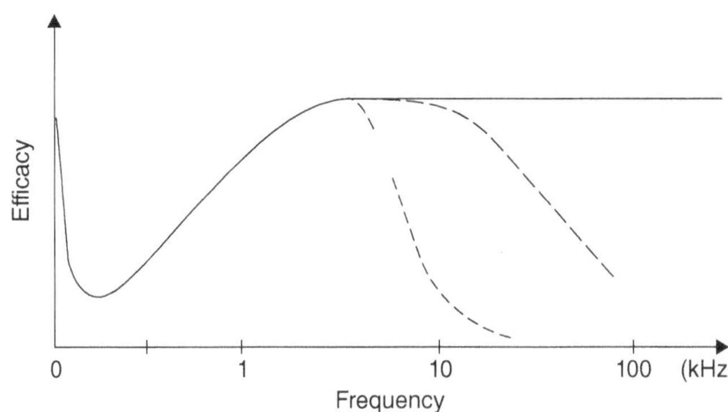

Figure 6.1. Frequency impact on efficacy of the system, as per the Sayyou Japanese patent (19991130) with what was shown afterwards through experimental results.

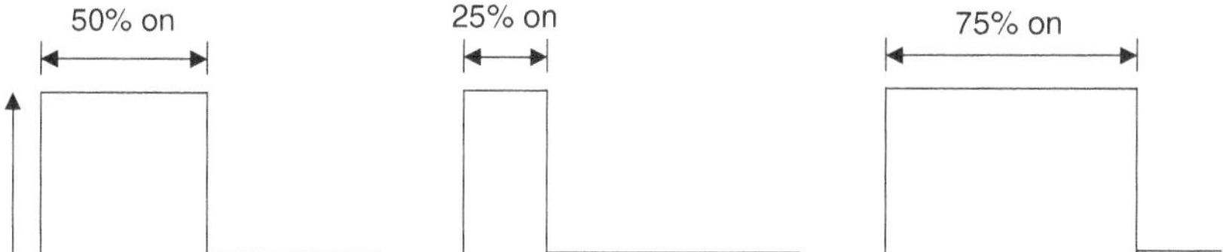

Fig. 6.2. Example of pulse width modulation (PWM) control.

average current can cause this, but so can other controls over the PWM, such as spark control, overcurrent protection shutdowns, etc.

This effect is greatly attenuated through the capacitor bank, which performs as an energy consumption buffer between the power source and the DC/AC converter. For short periods of time, the capacitor bank serves as a reserve of energy if consumption is momentarily greater than generation, and the other way around. The down side of this system is the high current flow into the capacitor bank to make the initial charge when starting or restarting the system.

Solution to high-energy consumption at the capacitor bank: Capacitor bank consumption attenuator

When starting or restarting the equipment, the current flow into the capacitor bank can be very high and demanding on the power generating components.

To attenuate this, it is necessary for a high-power and high-speed transistor (metal–oxide–semiconductor field-effect transmitter [MOSFET], insulated-gate buffer transistor [IGBT], etc.) to control the PWM in the very high current consumption moments. This transistor is to be controlled by the control system and provide a warranty that the power generating components will not suffer overcurrent issues.

The current control capacitor bank consumption attenuator (IGBT; which has never been used before for electrical invasive plant control) shown in Schematic 6.3 can be controlled dynamically from the output of an ammeter with a digital signal coupled with the control system or just become active whenever the system starts or restarts. In an idealized example of how much current can go through, once the capacitor bank consumption attenuator is at work, the frequency of modulation for the capacitor bank consumption attenuator band should be 1 kHz–1 MHz.

Multiple applicator segment through parallel current controlled transformer system

The first approach to multiple electrodes was in 2005 (Schwager and Schwager, 2005a, 2005b). Due to the large size of the traditional 50–65 Hz transformers, it is difficult to control the energy at a smaller segment, each of which would have to be connected to only one transformer. This causes the application quality to be uneven, especially when some plants provide a low electric resistance path, leaving other plants without enough energy for good quality control across the whole width of the applicator.

This technology alternates full power through each transformer and respective electrode in the system, "concentrating" the energy into a diminished number of plants with each commutation cycle, which should be between 0.01–2.0 s. This technology has the stated objective of diminishing the total amount of energy needed to electrocute a large number of plants simultaneously. Although the claim has practical value for 50–65 Hz systems, the technology greatly limits the potential number of transformers that can be used simultaneously. If there are x transformers, for it to be possible to commutate one transformer individually, there is a need to turn off $x-1$ transformers.

In a practical example of individually controlling 50 separate transformers (in the case of extensive soybean areas in Brazil), even when using the minimum amount of "on time" commutation for each transformer (0.01 s), it would take 0.5 s for the same electrode to be active again.

If said equipment were to travel at 5 km h^{-1} (1.4 m/s), which is a reasonably standard speed for agricultural equipment, the example's electrode

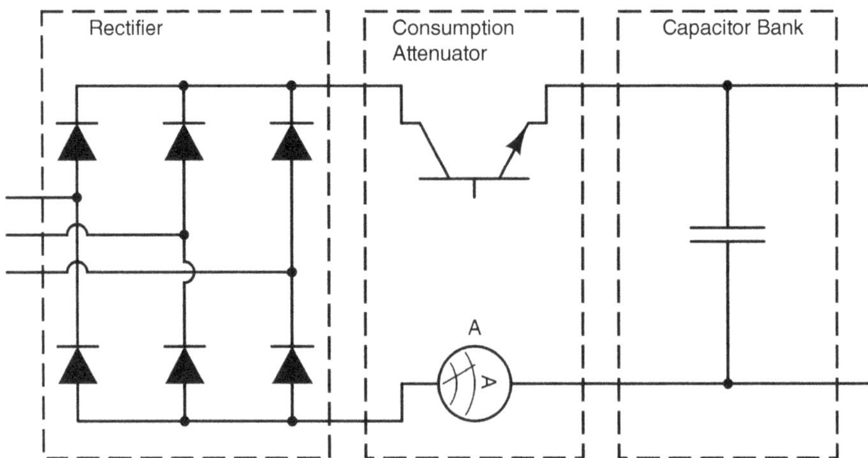

Schematic 6.3. Capacitor bank consumption attenuator.

segment would have gone through 0.7 m before being active again. As most invasive plants to be controlled or electrocuted are below 0.7 m in height, this method is unrealistic for larger numbers of parallel transformers/electrode segments.

Moreover, the system considered using a triphasic transformer, which needed an AC/DC converter to minimize the needs of electrodes for each transformer. A triphasic transformer would need three different points of contact (A, B, and C)—electrodes—which made it convenient for the AC/DC converter, since it diminished this number to two electrodes (+ and −).

A simplified circuit for the single commutation of one transformer using the technology described in PI 0502291, in the range of 2–100 Hz (0.01–2.0 s) is shown in Schematic 6.4:

Through a similar technology used for the capacitor bank consumption attenuator, the individual transformer PWM current control is proportionate to an individual PWM current control for each transformer and its correspondent electrode. The individual transformer PWM current control modulates the cycles themselves, since the frequency of the individual transformer PWM current control is to be smaller than the frequency of the system DC/AC converter (Schematic 6.5).

The idealized PWM format of the combined PWM from the DC/AC converter and the individual transformer PWM current control is shown in Fig. 6.3.

Unlike the capacitor bank consumption attenuator, which modulates DC through PWM and the technology described in PI 0502291—electronic commutation equipment for the electrocution of weeds (Schwager and Schwager, 2005a, 2005b), the modulation of which is in approximately the same frequency range as the main frequency (0.5–100 Hz and 50–65 Hz)—the individual transformer PWM current control modulates at the proposed range of 100 Hz–10 kHz. This frequency itself, although necessarily smaller than the DC/AC converter frequency, is in a much higher range.

Considering that, *ceteris paribus*, transformer size is inversely proportional to frequency, higher frequency DC/AC converters (1–18 kHz or above) allow for the use of much smaller transformers, making it feasible and possible to use a larger number of smaller power transformers.

The combination of the newly proposed technology (since 2018) and the use of higher frequencies DC/AC converters makes it possible to not only divide the power between the segments, but also to control the current at each segment individually with dynamic precision through the PWM control of the AC of the output of the DC/AC converter. Such a solution has never been used before for electrical invasive plant control.

Buck-boost voltage control

There are two main reasons for the adoption of this topology, which has never been used before for electrical invasive plant control (Schematic 6.6):

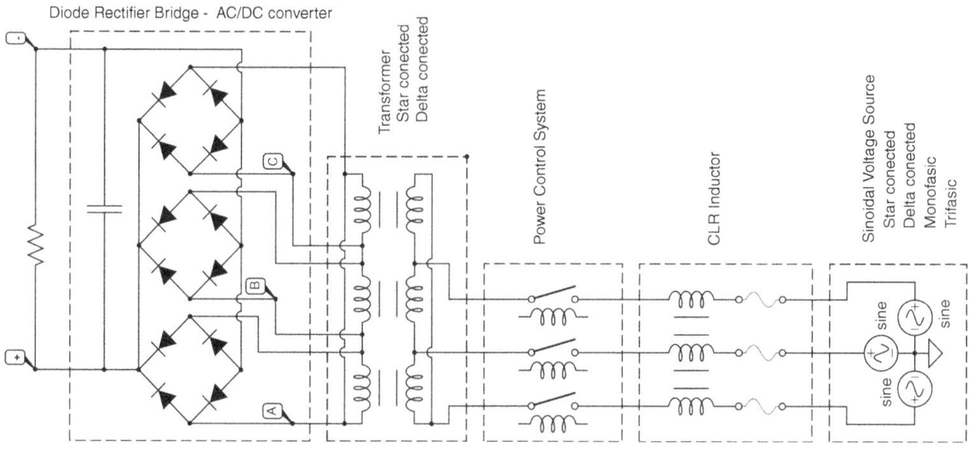

Schematic 6.4. P04437: individual transformer current control.

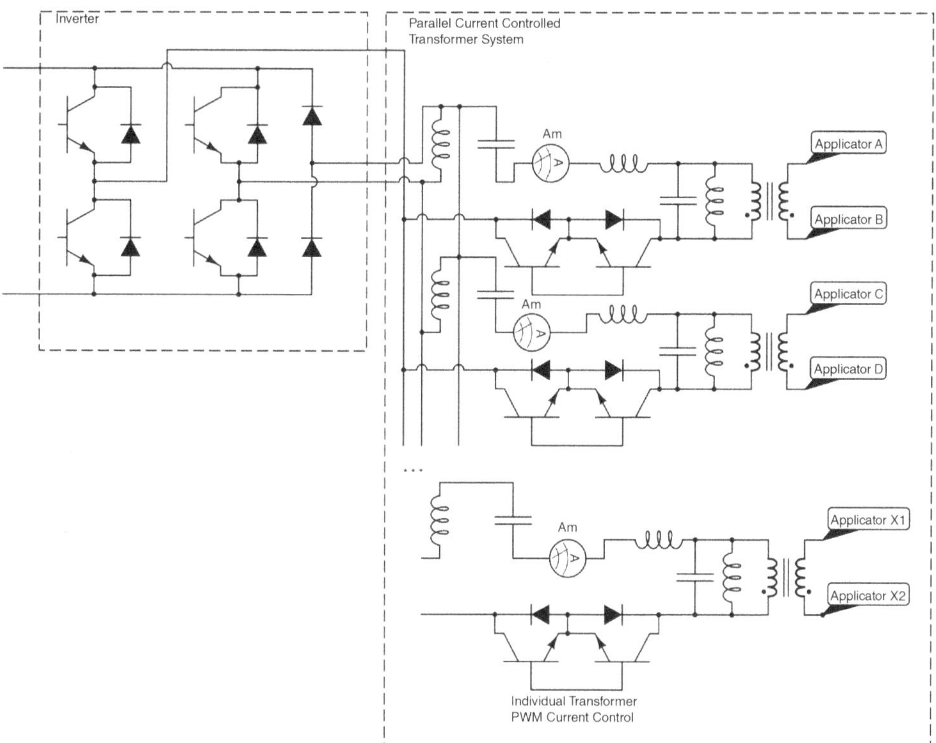

Schematic 6.5. Multiple applicator segment through a modular-paralleled current-controlled transformer system.

- Reduction of transistor stress.
- Dynamic voltage adjustment in the DC input of the DC/AC converters, adjusting voltage input at the primary of the transformer (therefore in the secondary as a result) to ensure the continuity of a stable power at the transformer secondary, even with a dynamic and rapidly changing electrical resistance (plant–soil electrical resistive system).

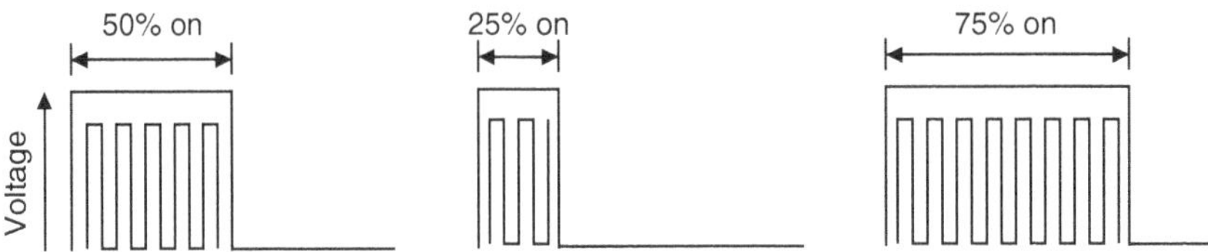

Fig. 6.3. Example of pulse density modulation (PDM) control.

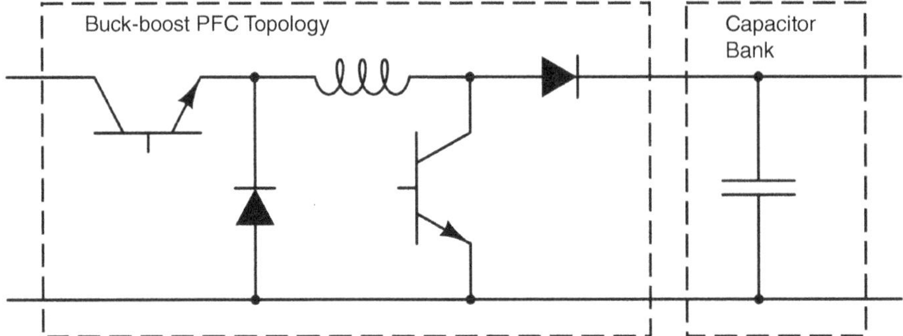

Schematic 6.6. Buck-boost voltage control.

The output load behaves like a variable resistance, so when the resistance value decreases, the secondary current increases proportionally, leading to the necessity of decreasing the secondary voltage, to keep the same power transfer. The control strategy will be developed to keep the system operating in these conditions. When the load current value changes, the transistors' commanding control strategy will change, modifying the converter output voltage (DC link inverter value), adapting its value to keep the power transferring to the load constant.

Harmonic voltage peaks limitation strategies

The use of square waves in high-frequency transformers is a common approach in inverter-based power systems due to its simplicity and ease of generation. Unlike sinusoidal waveforms, which require more complex circuitry to produce, square waves can be generated with minimal components, making them a cost-effective and efficient choice for high-voltage applications, such as electrical weeding systems.

However, the inherent characteristics of square waves introduce significant challenges, particularly when dealing with high resistance conditions in the plant–soil system. These challenges primarily arise from the harmonic components present in square waves and their impact on transformer behavior under varying load conditions.

A square wave is composed of a fundamental frequency along with an infinite series of odd harmonics (third, fifth, seventh, etc.). These harmonics do not cancel out, as they do in a pure sine wave, but instead create complex interactions within the electrical system.

In normal operation, a high-frequency transformer converts the inverter's AC square wave input into a high-voltage output that is then applied to the electrodes for weed control. However, due to the rich harmonic content, square waves can cause unintended resonance effects, high-frequency losses, and unexpected voltage spikes when certain conditions arise. When the secondary coil of a high-frequency transformer is lightly loaded or open-circuited, the voltage behavior can become unpredictable and unstable. This occurs due to:

- **Resonant overvoltage phenomena.** The interaction of high-frequency harmonics with the inductive and capacitive properties of the transformer windings can lead to excessive voltage amplification, creating hazardous conditions.

- **Voltage doubling effects.** At certain load conditions, standing waves can form, causing local voltage peaks that far exceed the transformer's rated output.
- **High-frequency eddy currents and core saturation.** The additional harmonics in the square wave can induce eddy currents within the transformer core, leading to heating, increased losses, and even potential failure.

These problems become especially severe in an electrical weeding system, where the load (plant–soil resistance) is highly variable and unpredictable. One of the most critical operating conditions for electrical weeding systems is the presence of high electrical resistance patches in the soil and plant structures. If the electrodes encounter an area of extremely high resistance (e.g. dry soil, isolated weeds, or weak electrical connections), the transformer's secondary side effectively behaves as if it is facing an open circuit. From the transformer's perspective, this is equivalent to running without a connected load, which exacerbates voltage instability and increases the risk of electrical overstress. The presence of harmonic-rich square waves further amplifies these effects, making it even more likely that dangerous voltage spikes will occur.

Given the inherent risks associated with square waves in high-frequency transformers, a voltage peak limitation strategy (VPLS) is an essential component for ensuring the safe and reliable operation of electrical weeding equipment. An effective VPLS should incorporate:

- **Overvoltage clamping circuits.** Using transient voltage suppressor (TVS) diodes, snubber circuits, or active crowbar protection to prevent excessive voltage spikes on the transformer secondary.
- **Adaptive load regulation.** Ensuring that the inverter dynamically adjusts output based on real-time load conditions to prevent open-circuit scenarios.
- **Resonant suppression filters.** Employing inductor-capacitor filter networks to dampen the effects of harmonics and minimize resonance-induced voltage amplification.
- **Transformer core optimization.** Using ferrite or nanocrystalline cores with optimized winding configurations to reduce eddy current losses and enhance performance under variable load conditions.

While square waves are among the simplest waveforms to generate, their use in high-frequency electrical weeding transformers requires careful consideration due to their potential to cause (Fig. 6.4):

- harmonic-induced voltage spikes;
- resonance effects in open-circuit conditions;
- transformer overheating and electrical stress.

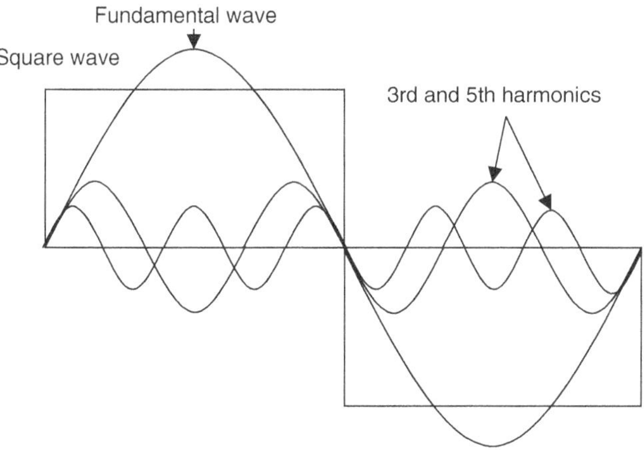

Fig. 6.4. Square wave decomposition in harmonics.

To safely integrate square waves into electrical weeding systems, a robust VPLS is mandatory to mitigate risks, ensure system reliability, and protect both the equipment and operators from unexpected voltage surges. The harmonic components can interact with each other if there is no load in the secondary when using a square-wave input. This interaction can cause large voltage peaks and undesired behavior.

Figure 6.5 shows an actual voltage wave at a lab test when the load disappears. The voltage peaks reach 10 kV+, from a previous 5 kV- secondary coil transformer:

External inductor

If the transformer construction of the system has small leakage inductance, an external inductor can be added in series with the transformer and used to clamp the voltage peaks to a maximum equal to the DC link. This solution has the advantage of being very robust, but the disadvantage of requiring a low magnetic dispersion transformer to work properly.

A simplified circuit design is shown in Schematic 6.7.

Figure 6.6 shows a simulation for a given transformer of the voltage wave format without (before) and with (after) this protection system (first wave has load, the three subsequent waves are in open circuit):

Secondary parallel alternating current/direct current converter solution

To avoid the harmonic voltage peaks, a rectifier can be added in parallel with the electrodes after the secondary coil of the transformer (Schematic 6.8). This solution allows for very good and precise maximum voltage limitation, but has the disadvantage of requiring (usually expensive) fast and powerful diodes for high-power, high-frequency systems.

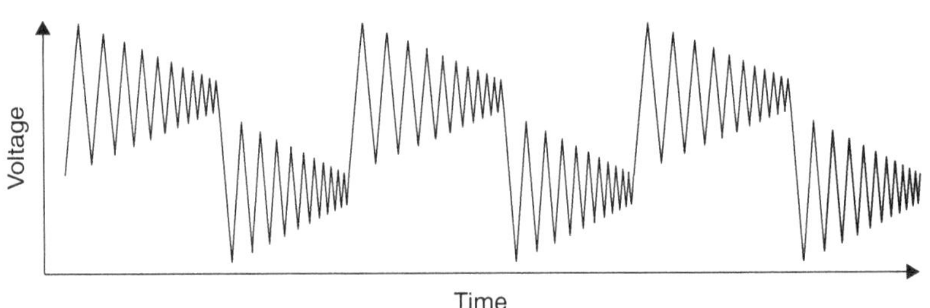

Fig. 6.5. Voltage spikes where none of the voltage peak limitation strategies (VPLS) have been used.

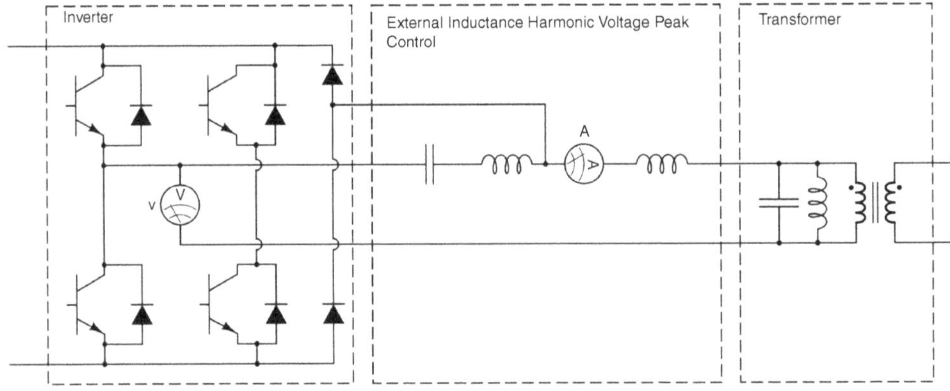

Schematic 6.7. External inductor schematic for voltage peak control.

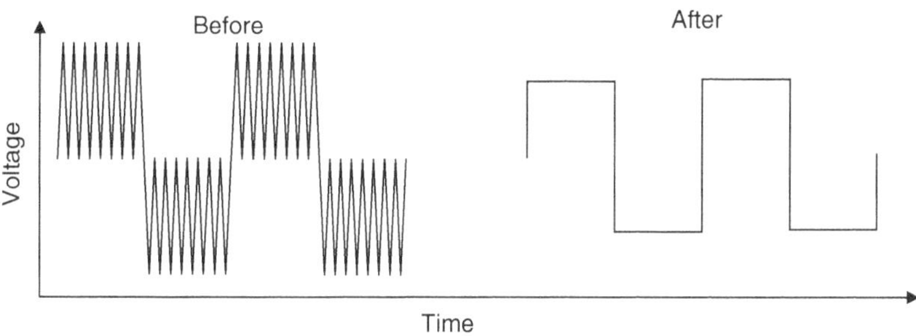

Fig. 6.6. Before and after the implementation of the voltage peak control.

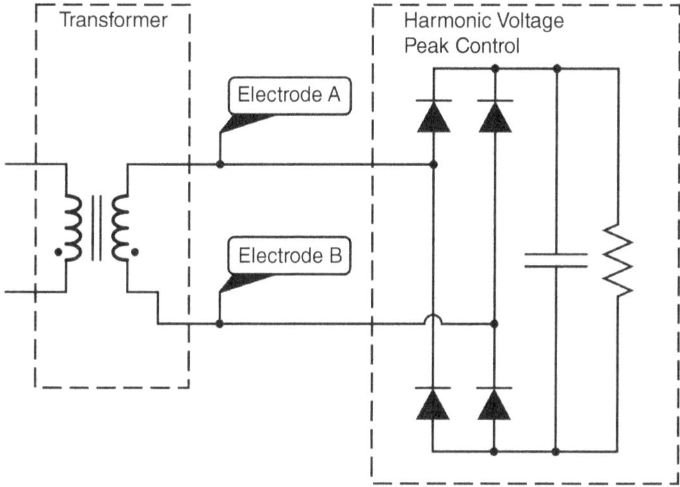

Schematic 6.8. Secondary parallel alternating current (AC)/direct current (DC) converter solution.

Quasi-square wave "Harmonic Killer"

The strategy used in this case is to take out the harmonic component from the primary voltage, making a quasi-square wave. This will avoid the resonance excitation and consequently the overvoltage peak (Fig. 6.7). The wave has a step of zero voltage that matches the most problematic harmonic component (Sankaran, 2002).

The advantage is that this solution needs no added hardware, but the disadvantage is that it provides a lower-quality maximum voltage limitation and, depending on the time and extension of the zero-voltage step in each wave, it might harm the total power a fix transformer can deliver and the power density of the system.

The use and topography of high-frequency transformers

The new and innovative circuits, software, and the development of new materials (e.g. silicon carbide for semiconductors and crystal ferrite for permeable magnetics) allow for the use of a high-performance high-frequency transformer (HFT). A usual ratio of the technology developed could be 1 kg/kVA (0.1 l/kVA volume density), already considering inverter weight. This is a one order of magnitude reduction from the 10 kg/kVA power density of traditional 50–65 Hz transformers.

- The performance of the system reduces as frequency increases, because heat dissipation

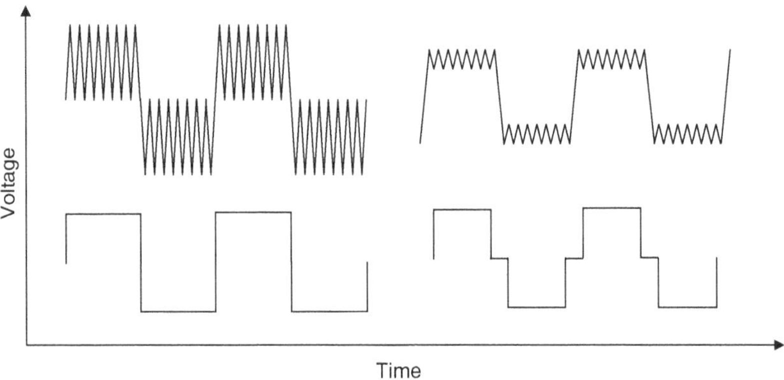

Fig. 6.7. Square wave (left) and quasi-square wave (right) input and output voltage comparison.

increases with higher frequencies at the inverter—which is undesirable.

- The size and cost of the transformer decreases as frequency increases (up to a point)—which is desirable.
- Transformer size and cost reduces as frequency increases with diminished marginal influence. This is caused because of the set area reserved by insulating materials, which cannot be reduced with higher frequencies and the skin effect, which increases the volume of the wire needed with higher frequencies. An alternative, to reduce the increase of the volume of the wire needed, is to use litz wire, but the reduction has a limit since each wire of the composition must be individually insulated with space-consuming coatings.

Because of these two facts, there is an optimum range of frequencies where it is possible to combine low weight, low cost, and high efficiency. With the usage of silicon carbide (or similar materials), IGBT inverters, and crystal ferrite magnetic permeable transformer cores, this optimum range is 15–35 kHz, being reasonable values varying from 1–100 kHz. Each transformer topology has a different cost–benefit relation to different frequencies, but all possibilities put together create, not an optimum frequency, but an optimum range of 15–35 kHz.

An example of how this relation behaves at a given fix HFT, is given in the "Volume/weight/cost comparison of a 1 MVA 10 kV/400 V solid-state against a conventional low-frequency distribution transformer" paper from Jonas E. Huber and Johann W. Kolar, from the Power Electronic Systems Laboratory, ETH Zurich in Switzerland.

Although composed only of coil/reel system, wire, core, connectors, and encapsulation, a high-frequency, high-power transformer construction is state of the art technology. In this same construction, you have to take into consideration a large array of interrelated variables that do not interfere with each other in a linear manner.

The individual parts and the relation between them must take into consideration variables such as:

- skin effect losses
- internal capacitance
- winding ratio
- thermal dispersion
- magnetic dispersion
- core magnetic permeability
- core physical resistance
- potential corona effects
- winding window optimum size
- core shape
- serial harmonics
- parallel harmonics
- high-voltage insulation in small spaces
- layer division
- resin viscosity

To diminish magnetic dispersion, the coupling between primary and secondary must be very high and precise. Traditional coils are wound one on top of the other, with protective insulating papers between layers, and this construction promotes little magnetic dispersion. In the HFT that is not possible because it generates too much internal capacitance, which therefore generates internal current, heat, and power loss.

Vertical slots are used in some high-voltage transformers, but in the HFT the number of layers needed is not viable, because of the combination of small size and high tension. Therefore, the solution is to combine intercalated 3–7 secondary-primary-secondary organized slots with insulating paper between them. This seemingly simple solution guarantees low magnetic dispersion with low internal capacitance. Such a solution has never been used before for electrical invasive plant control transformer construction.

Higher frequencies increase skin effect, causing current to flow disproportionally through the outer part of the wire, instead of evenly. Because of this, the wire must be litz-organized. Litz wire consists of multiple strands insulated electrically from each other.

This fact also limits frequency since for higher frequencies a larger number of wires would be needed and that reduces the ratio of conducting/insulating material, since smaller wire gauges have more insulation for the same amount of conductor (the wires composing the litz must be insulated from each other, otherwise they would behave as just one wire). Infinite frequency would require an infinite number of insulating materials to insulate the wires in the litz organization.

As an example of the effect this has at higher frequencies, in Table 6.1 there is representative parameter data for a 24-gauge plastic-insulated conductor (PIC) telephone cable at 21°C (70°F).

This seemingly simple solution (litz wires) is key for the use of HFTs, since it greatly increases the surface area of an orthogonal cut, therefore increasing the maximum current that can go through a high-frequency wire of a set radius.

Usually, electronic vacuum encapsulation happens using a simple pouring in container method (Fig. 6.8). This is easy and cheap to do, but presents two issues with the HFT:

- It creates a large resin layer that thermally insulates the transformer, causing it to unavoidably overheat over time;

Table 6.1. Frequency effects on transformer wiring.

Frequency (Hz)	R (Ω/km)	L (mH/km)	G (μS/km)	C (nF/km)
1	172.24	0.6129	0.000	51.57
1k	172.28	0.6125	0.072	51.57
10k	172.70	0.6099	0.531	51.57
100k	191.63	0.5807	3.327	51.57
1M	463.59	0.5062	29.111	51.57
2M	643.14	0.4862	53.205	51.57
5M	999.41	0.4675	118.074	51.57

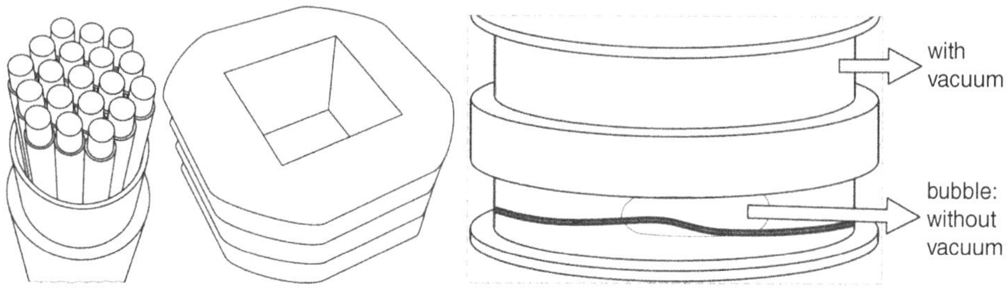

Fig. 6.8. Wire insulation and encapsulation, transformer slotted reel structure example, and vacuum effect in encapsulation.

- It does not penetrate deeper layers, because of the small size and closeness of the components.

Because of those facts, a vacuum bag is used, where the resin remains in normal air pressure.

The low viscosity resin entrance must be at the lowest central point of the reel and the vacuum suction at the highest central point to ensure the resin goes through all of the transformer volume before reaching the air exit point. The flow goes against gravity to ensure there are no air bubbles. This process ensures there is not only no air inside the transformer, but also no empty space—both of which could contribute to corona effects and insulation problems. Bags and forms can be used. Bags have the advantage of no extra resin being necessary and forms have the advantage of being reused multiple times.

Such a solution has never been used before for electrical invasive plant control transformer construction.

Alternating current vs direct current: The power factor

Power delivery and efficiency considerations in alternating current vs direct current for Generation 2

Generation 2 electrical weeding systems can operate using either AC or DC. However, when implementing an AC-based system, it is crucial to account for the inherent inefficiencies associated with AC power delivery, compared to DC. These inefficiencies impact both power transfer efficiency and electrical efficacy, ultimately influencing the effectiveness of the weed control process (Coutinho Filho *et al.*, 2017).

Understanding the power factor and its role in alternating current efficiency

One of the most significant factors affecting the efficiency of AC power systems is the power factor (λ).

The power factor (denoted as λ) is defined as the ratio of real power (active power) to apparent power in an AC circuit. It can also be described as the absolute value of the cosine of the phase angle (φ) between the voltage and current waveforms (Eqn 6.1).

$$Power\ Factor(\lambda) = \frac{Active\ Power}{Apparent\ Power} = \frac{VI * \cos\phi}{VI} = \cos\phi$$

Equation 6.1. The power factor.

where:

- **Active Power (P)** (measured in watts) is the actual power consumed by the load to perform useful work.
- **Apparent Power (S)** (measured in volt-amperes) is the total power flowing in the circuit, including both active power and reactive power.
- **Reactive Power (Q)** is the power stored and released by inductive and capacitive elements in the system, contributing to inefficiencies.

Since the power factor is always less than 1 in AC systems ($\lambda<1$), it is a key factor limiting the efficiency of AC power transfer.

Power factor implications in alternating current vs direct current systems

Unlike DC systems, where voltage and current are always in phase (power factor = 1), AC systems suffer from phase shifts due to inductive and capacitive components in the circuit. These elements lead to:

- **Increased apparent power**. The total power supplied by the source must be higher than the actual power required by the load due to the presence of reactive components.
- **Lower energy transfer efficiency**. A lower power factor means that a significant portion of energy is oscillating between the source and the load, rather than being converted into useful work.
- **Heat generation and system losses**. The presence of inductive and capacitive reactance in AC circuits results in additional heat dissipation and increased energy losses, requiring larger power ratings to achieve the same effective output as a DC system.

These factors directly reduce the electrical efficiency of an AC-based system, meaning less energy is effectively used to kill weeds compared to a DC-based system operating at the same apparent power level.

Practical consequences for electrical weeding systems

When choosing between AC and DC for electrical weeding systems, the following performance trade-offs must be considered:

ALTERNATING CURRENT SYSTEMS

- Can leverage high-frequency effects to enhance certain biological interactions with plant cells.
- Lower power efficiency due to power factor limitations ($\lambda<1$).
- More complex impedance interactions with soil, plant tissue, and electrodes.
- Higher risk of energy losses due to reactive power circulation.
- Higher safety risks due to higher voltage peaks and other implications.

DIRECT CURRENT SYSTEMS

- 100% power efficiency in energy delivery ($\lambda=1$), meaning all supplied power contributes to plant destruction.
- Simpler circuit behavior with direct resistive heating effects in plants.
- More predictable and stable electrical penetration into root structures.
- Lack of frequency-dependent effects that might be useful in specialized applications.

Conclusion: Direct current is generally more efficient for electrical weeding

Given the importance of power efficiency in electrical weeding, DC-based systems generally offer superior performance compared to AC systems due to their higher power transfer efficiency, minimal energy losses, and more predictable interaction with plants and soil:

- AC power factor limitations ($\lambda<1$) result in reduced overall energy effectiveness, leading to lower electrical efficacy in plant control.
- DC systems, with $\lambda=1$, ensure that nearly all delivered power is effectively utilized, making them the preferred choice for maximizing weed destruction efficiency.

While AC systems might be beneficial in niche applications where frequency effects are desirable, for general-purpose electrical weeding, DC remains the optimal choice due to its superior power efficiency and direct energy transfer characteristics.

Generation 3

The generations relate to weed inactivation devices, comprising at least two electrodes, where at least one electrode is directed to the weed. The weed activation device is used as a physical herbicide apparatus. A weed activation device of the generic type has been disclosed in Gen. 2, where high voltage is created by the utilization of high-voltage transformers. The high voltage is applied to an electrode which contacts the weed to be controlled or is brought close to the weed. This physical herbicide has the great advantage of not utilizing chemical herbicides, which may proliferate into the food chain up to humans.

The object of the current invention is to provide for an electronic topology or circuitry, which allows the use of small, cheap, and available electronic components to comprise a high power factor converter that controls for power without the need of software, or other hardware components required in previous technological generations. This particular converter is composed of, at least, the following components:

- an inverter;
- an inductive and/or capacitive harmonic filter;
- a capacitive voltage multiplier composed of diode(s) and capacitor(s).

A DC or AC power supply of any number of phases generates a voltage. This voltage is fed to an inverter that increases frequency. The current with increased frequency is fed to a harmonic filter (inductive, capacitive, or both) that ensures a high power factor, diminishing or excluding the need of a separated power factor converter (PFC). The inductive and/or capacitive harmonic filter may feed an HFT that further increases the voltage input for the voltage multiplier. The HFT may comprise a centered tap in its secondary winding, which can serve as a voltage reference or grounding to the secondary coil.

The output of the previous components is fed to a voltage multiplier (e.g. a voltage multiplier of the Cockroft-Walton type, or a full wave Cockroft-Walton type, such as a hexuplicator, multiplying the input voltage by a factor of six). The voltage multiplier provides different voltage levels depending on its load, so for a variable load it makes an auto-adjustable power control without the need for any additional circuitry, a processor, or a controller (Schematic 6.9).

If, as described, a transformer was necessary to further increase the voltage between the harmonic filter and the voltage multiplier, the voltage multiplier reduces the high-voltage levels necessary at the transformer secondary side, facilitating its construction and reducing its weight, volume, and insulation breakage risks. Also, the voltage multiplier always represents a series impedance connected at the transformer secondary, not letting the transformer in a direct real open circuit situation, and reducing the risks of series resonance excitation and voltage peaks that could damage insulation or create other internal damage.

This particular construction allows for the inverter switching to be set as resonant or quasi-resonant. This setting of the inverter as resonant or quasi-resonant reduces its output harmonic composition, reducing the risk of transformer series resonance excitation and, consequently, reducing the risk of compromise of the transformer insulation. Also, the inverter's switches (such as, but not limited to IGBTs, power transistors, and MOSFETs) have reduced conduction losses when working in the resonant or quasi-resonant mode, increasing the converter's overall efficiency.

Moreover, if a transformer was necessary to further increase the voltage between the harmonic filter and the voltage multiplier, the inverters resonance frequency is tuned as the resonance frequency between an inductor connected in series with the inverter and the total capacitive effect reflected to the transformer primary side.

In this example, a monophasic power supply is shown (Schematic 6.9), which is rectified and then fed into a half-bridge inverter. In this case, the voltage multiplier is a simple voltage doubler, and no filter is added.

A high-order multiplicator simplifies the transformer's construction even more, using the same justifications as before.

The external inductor reflected to the transformer's secondary side will also provide an impedance matching with the voltage multiplier series impedance, and this association with the "plant resistance" will be seen by the transformer as a resistance in parallel with a capacitor inversely proportional to this resistance value. The output voltage will be variable with the resistive load, since the voltage multiplier capacitor charging will be "controlled" by the total series impedance, such that different voltage values will be delivered by the converter, depending on the resistive load, according to the basic power equation P = V^2/R, where R, V, and P are the resistive load, its voltage, and its power dissipation, respectively. As the impedance matching happens in a self-adjustable way, this converter topology presents a self-adjustable power control without the necessity of any extra voltage, current, or power control strategy implementation.

The impedance seen by the transformer is still a resistance in parallel with a variable capacitor, as described before. The inverter's resonance switching can be tuned so the converter delivers the optimum maximum power to a specific impedance value. Also, the voltage multiplier series impedance partially solves the problem of transformer series resonance excitation, once the transformer's secondary is never in a real open-circuit situation with this new topology.

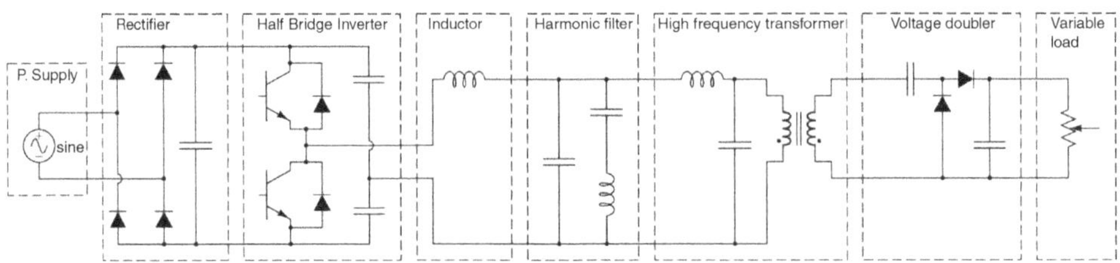

Schematic 6.9. Generation 3 simplified.

When the resistive load tends to a low value (a short-circuit situation), the voltage multiplier presents a series impedance reflected to the primary that, associated with the external inductor, protects the transformer against high short-circuit currents. When the load tends to a high value (open-circuit situation), all the capacitors of the voltage multiplier are charged, increasing the secondary voltage peak, but still limiting it to a maximum value equal to the multiplier stage.

The strategy for protecting the inverter and potentially also the transformer against dangerous operation parameters, is the addition of an adequate capacitive or capacitive inductive filter after the external inductor. As the filters can be projected to remove the high-order harmonic components from the transformer's input voltage, the series resonance excitation can be avoided with this strategy.

As the "plant resistance" deviates from the tuned value, the delivered power decreases from its optimum maximum value, but a considerable range of power values is still delivered to a great variety of resistive loads, as can be seen in Table 6.2, that show the power delivered to different values of resistive loads, considering power grid as power supply and a capacitive voltage hexuplicator. It's important to notice that the electronic converter as described was never used before for invasive plant control.

Another example of a more complete different composition of the proposed system, may show a three-phasic rectified power supply, with a full-bridge inverter feeding a harmonic filter with a transformer with centered tap feeding a double-voltage doubler as a voltage multiplier.

The state of the art for electrical weeding systems relates to the fact that weed pressure is a function of the density of weeds per area, the organic mass per area, and the type of weeds. The weed pressure faced by the equipment can vary, changing the energy needed to ensure proper weed control. As the ideal equipment should always be able to output constant power to maximize the use of the power source constraints, in the case of any surplus energy, due to low weed pressure, it is more efficient to adjust power delivery by means of a sensor that measures weed pressure, or to just speed up the equipment, reducing the energy per area.

Traditional systems that have AC or greatly varying voltages with fixed waveforms with high voltage peaks, such as the ones exemplified in patents WO2013051276, WO2015119523, WO2016016627 and EP3744173, have a hugely varying power delivery through changing resistances that will be suboptimal for their energy-efficiency usage. This is due to hugely varying energy consumption at the plant, leading to either an uneven quality of application or the need for a much larger power input capability, with higher energy consumption, to ensure

Table 6.2. Power curve attained with a resistive load range (2.77 to 100 kΩ) from Zasso's patent.

Example attained with power variation with the resistive load (2.77 a 100 kΩ)				
Power grid voltage value	220 V		127 V	
Resistive load (kΩ)	Current root mean square (RMS) (A)	Active power (W)	Current RMS (A)	Active power (W)
100	1.25	270	1.3	165.1
75	1.55	334.8	1.64	208.28
50	2.2	475.2	2.29	290.83
25	3.05	658.8	3.33	422.91
12.5	2.23	481.68	2.55	323.85
8.5	1.75	378	2.04	259.08
6.25	1.47	317.52	1.7	215.9
5	1.09	235.44	1.46	185.42
3.57	0.929	200.664	1.18	149.86
2.77	0.838	181.008	1	127

enough energy is delivered when resistances are high because plants, plants–soil, or plants–soil–plants resistive systems vary greatly in practical applications. It is possible to even the average power delivered through continuity switches and controls, such as PWM and pulse density modulation (PDM), as thought in EP3744173, but this accounts for average power, not continuous power delivery, which can only be achieved by a continuous DC that varies its voltage to account for dynamic resistance changes to ensure continuous power delivery. A brief history of attempts that are improved upon with the latest technology are:

- Patent EP3744173 (Coutinho Filho *et al.*, 2020) explains thoroughly the use of PWM and PDM to control for power in electrical weeding operations. It offers a constant voltage architecture as the output of a transformer, and the PWM and PDM controls work as a continuity switch that controls the power by controlling the continuity of the current. In other words, it is a constant AC voltage supply that is controlled by continuity to limit current, offering a constant power source. This is a sub-optimal solution considering the output is not constant DC that only varies with the dynamically changing loads to ensure constant power delivery.
- Patent EP3557750 (Valverde *et al.*, 2019) describes a frequency converter, a transformer, and a capacitive voltage multiplier composed of diodes and capacitors. To ensure semiconstant power, it uses the impedance matching implicit of the voltage multiplier, which happens in a self-adjustable way, without the necessity of a control strategy implementation. This happens, to a certain degree, because when the resistive load tends to a low value (short-circuit situation), the voltage multiplier presents a series impedance reflected to the primary that, associated with the external inductor of the filter, protects the transformer against high short-circuit currents. When the load tends to a high value (open-circuit situation), all the capacitors of the voltage multiplier are charged, increasing the secondary voltage peak, but still limiting it to a maximum value equaling the multiplier stage. In other words, when the load is higher than the load that delivers maximum power, the power diminishes because the converter cannot increase the voltage enough to ensure constant power, and when the load is lower than the load that delivers maximum power, the internal impedance reflected increases, lowering the power delivered.
- PCT WO2013051276 (Yuasa and Yuasa, 2013) contains a frequency converter, a transformer and a capacitive voltage multiplier composed of diodes and capacitors, more specifically a Cockroft-Walton circuit for several uses. In this invention, the concept is to ensure a voltage boost without impedance matching, multiple peaks of voltage, constant or semiconstant power, or PWM controls. As proposed, a constant power delivery is optimal for electrical weeding, therefore, it may be used, but it would not fit all scenarios, especially the ones where the load level is not the one where the system outputs peak power or the load varies dynamically, which is the operational scenario of a regular electrical weeding operation. Therefore, it is not suited to the target operation of the present invention.
- PCT WO2015119523 (Stanković, 2015) contains a feedback assembly of transformers and voltage multipliers. It solves the challenge of the dynamically changing load that electrical weeding devices face through a feedback loop that feeds the final transformer outputting AC or highly varying voltage output with a constant voltage waveform. Output coming from a transformer directly means that this varying voltage will have a power curve in time that is not constant, therefore its output cannot provide the ideal constant DC that varies its voltage according to the output load ensuring constant power delivery. Therefore, although it can vary its voltage to face a dynamically changing load, it cannot provide the optimum voltage constantly through a constant DC that varies its voltage according to the output load ensuring constant power delivery. It may provide some degree of impedance matching or some degree of semiconstant power, but as high peaks are parasites to the optimum voltage and constant

power delivery, it cannot provide an ample range of semiconstant power for a given range of loads, nor can it offer a constant DC output that only varies according to the load, keeping the power output constant. For those reasons, the invention described in WO2015119523 is not optimal to the target objective of the present invention.

- PCT WO2016016627 (Diprose *et al.*, 2016) relates to the use of outputting processed electrical energy that comprises a waveform with a frequency of at least 18 kHz or more, with peak voltages of at least 1 kV. Although this invention may output for a functioning piece of equipment, with the advantages stated in the referred document, it is suboptimal. As described, the invention of WO2016016627 cannot deal with the varying impedances to output constant power (neither can semiconstant power, for that matter). Moreover, the voltage peaks of more than 1 kV reduce the efficacy of the use of electrical energy by not delivering constant DC that only varies with the dynamically changing loads to ensure constant power delivery. WO2016016627 proposes that the use of higher frequency is less dangerous to humans, but this effect is offset by the need for a much higher voltage output, because to achieve the same power a system with high voltage peaks will need higher peak voltage than a system with continuous DC. This effect is much increased when it is taken into consideration that the power capabilities of such a system will have to be even higher to account for varying impedances, even if it has a PWM system to ensure that average power is constant. Average power being constant does not mean peak power is constant or closely so.

Although it has some degree of power control, the solutions of the prior art are semiconstant, given the nature of the power output curve of a regular voltage multiplier. To improve upon this, a multiple setting of different values for the voltage multiplier can be constructed. For instance, paralleling multiple voltage multipliers with different peaks may provide a power curve with multiple peaks, where the power output is stable at a much wider load range. The results of a combination of two and three of these voltage multipliers provide multiple peaks where a growing stable area of semiconstant power can be found.

To achieve the objective to provide a small, cheap, high-power factor efficient and effective electronic converter to control for constant power output with minimal voltage parasitic fluctuations, the technology, as proposed in Rona *et al.* (2019), is a weed inactivation device comprising an electrical power source. It has at least one electronic converter comprising: a power inverter; a transformer coupled to the inverter; a voltage multiplier (impedance matcher coupled to the transformer); a plurality of electrodes coupled to the at least one electronic converter, wherein at least one electrode is pointed at one or more loads; and a control unit comprising a PWM module and at least one sensor, wherein the PWM module is coupled to at least one electronic converter and at least one sensor is coupled to the plurality of electrodes. The control unit controls the power output of the at least one electronic converter to supply, within a determined load range, substantially constant power to the plurality of electrodes by adjusting the duty cycle of the PWM module according to the load; and limiting parasitic voltage peaks to 1 kV.

References

Coutinho Filho, S.A., Pomilio, J.A., Valverde, B. and Souza, D.T.M. de (2017) *Dispositivo de Desativação de Ervas Daninhas. PCT/IB2017/001456*. World Intellectual Property Organization (WIPO), Geneva, Switzerland.

Coutinho Filho, S.A., Pomilio, J.A., Valverde, B. and Mendes de Souza, D.T. (2020) *Weed Inactivation Device. Patent No. EP3744173*. European Patent Office (EPO), Munich, Germany.

Diprose, M.F., Coleman, G.H. and Diprose, A.J. (2016) *Apparatus and Method for Electrically Killing Plants. Patent No. WO 2016/016627 A1*. World Intellectual Property Organization (WIPO), Geneva, Switzerland.

Rona, S.A., Valverde, B., Souza, D.T.M. de and Coutinho Filho, S.A. (2019) *Dispositivo de Inativação de Plantas Invasoras. Patent No. BR 10 2019 002353-8*. Instituto Nacional da Propriedade Industrial (INPI), Rio de Janeiro, Brazil.

Sankaran, C. (2002) *Power Quality, 1st edn*. CRC Press, Boca Raton, USA.

Sayyou Co., Ltd. (1995) *Superconducting-type Weed Extermination Device. Patent No. JP H07-89 A*. Japan Patent Office (JPO), Tokyo, Japan.

Schwager, A.H. and Schwager, J.R. (2005a) *Equipamento de Comutação Eletrônica de Eletrodos Múltiplos para Eletrocussão de Plantas Daninhas. Patent Application PI0502291-6 A2*. Sayyou Brasil Indústria e Comércio Ltda, São Paulo, Brazil.

Schwager, A.H. and Schwager, J.R. (2005b) *Equipamento de Comutação Eletrônica de Eletrodos Múltiplos para Eletrocussão de Plantas Daninhas. Patent No. PI 0502291-6*. Instituto Nacional da Propriedade Industrial (INPI), Rio de Janeiro, Brazil.

Stanković, M. (2015) *Apparatus for Destroying Weeds. Patent No. WO 2015/119523 A1*. World Intellectual Property Organization (WIPO), Geneva, Switzerland.

Valverde, B., Mendes de Souza, D.T., Coutinho Filho, S.A. de and Rona, S.A. (2019) *Weed Inactivation Device. Patent No. EP 3 557 750 A1*. European Patent Office (EPO), Munich, Germany.

Yuasa, Y. and Yuasa, M. (2013) *The High-Voltage Application Device. Patent No. WO 2013/051276 A1*. World Intellectual Property Organization (WIPO), Geneva, Switzerland.

7

Ecotoxicology

Introduction

As a physical weed control process, electrical weeding does not fall under the regulatory scope of European Union chemical registration laws, such as those governing herbicides and chemical pesticides. Unlike chemical weed control methods, which require extensive toxicity and environmental safety approvals, electrical weeding operates through direct energy transfer rather than introducing synthetic compounds into ecosystems. However, despite the absence of regulatory obligations, Zasso has taken a proactive approach by conducting extensive ecotoxicological studies in collaboration with professional laboratories. These studies aim to ensure environmental responsibility and demonstrate the long-term sustainability of electrical weeding as a viable alternative to chemical herbicides.

Recognizing the importance of soil ecosystem integrity, predictive ecotoxicological studies to assess the potential side effects of electrical weeding on soil-dwelling organisms were conducted by Zasso and shared. These investigations have focused on key bioindicators, such as earthworms, microorganisms, and other soil invertebrates, which play a crucial role in nutrient cycling, soil structure maintenance, and plant health. Field trials were designed to differentiate the impacts on epigeic (surface-dwelling) and endogeic (subsurface) soil fauna, ensuring a comprehensive evaluation of biological effects.

Field trial methodology and evaluation

To assess the potential ecotoxicological impact of electrical weeding, standardized field trials were conducted in different soil conditions and climatic regions. The studies analyzed:

- **Soil fauna activity and survival rates.** Earthworm populations and other invertebrates were monitored before and after treatment, evaluating potential impacts on reproduction, mobility, and survival.
- **Microbial communities and soil respiration.** The impact of electrical weeding on microbial diversity was assessed using soil respiration tests, microbial biomass measurements, and enzyme activity assays to determine if beneficial bacteria and fungi were affected.
- **Comparison with conventional methods.** The effects of electrical weeding were compared to mechanical and chemical weed control methods, evaluating differences in soil health impact and biodiversity preservation.

DOI: 10.1079/9781836992288.0007

Independent research results and environmental sustainability

The independent ecotoxicological research confirmed that electrical weeding is a sustainable and environmentally friendly method of weed control. Key findings from these studies include:

- **Minimal impact on soil fauna.** Short-term exposure to electrical discharge did not cause significant mortality in earthworms or other soil organisms. While temporary behavioral responses (such as retreating deeper into the soil) were observed, populations fully recovered within standard observation periods.
- **No lasting effects on microbial communities.** The soil microbial diversity and activity remained stable after treatment, with no measurable reduction in beneficial microbial populations.
- **No residual soil contamination.** Unlike chemical herbicides, electrical weeding leaves no chemical residues in the soil, ensuring that subsequent crops and beneficial organisms are not negatively affected.
- **Preservation of soil structure and fertility.** Unlike mechanical weeding, which can disturb soil layers and lead to erosion or compaction, electrical weeding minimizes soil disruption, preserving natural soil processes and nutrient cycles.

The results of these independent studies reinforce electrical weeding as an eco-friendly alternative that effectively controls weeds without harming soil health, biodiversity, or ecosystem stability. As agriculture increasingly shifts toward chemical-free weed control solutions, electrical weeding stands out as a pioneering innovation, combining efficiency, safety, and sustainability for the future of environmentally responsible farming.

Ecotoxicological Study on Grassland (by BioChem agrar GmbH, Germany)

An initial study was commissioned in 2019 to investigate effects on springtails (Collembola) and earthworms (Lumbricidae) at the population level on undisturbed grassland, with high abundances of soil organisms (springtails ~13,445±3,295 individuals/m^2; earthworms ~310±42 individuals/m^2; Fig. 7.1). The sampling was performed by a German contract laboratory (BioChem agrar GmbH) according to ISO 23611-2 (2006) as well as ISO 11268-3 (2014). Samples were taken before (T0) and four weeks after (T1) Electroherb™ applications (System: Xpower; 3 km h^{-1}; ~80 kWh ha^{-1}). The test followed a randomized design with two repetitions of the treatment and 2–3 soil core samplings per repetition (n≤4). The abundances at the second sampling after treatment with Electroherb™ showed that when applied under realistic field conditions with energy dosages found to be effective for weed control, no or only minor effects on soil organisms were observed.

The results presented indicate, that Electroherb™ did not severely influence the abundance of springtail populations at the site four weeks after the treatment. Neither of the Electroherb™ application times ("in the morning" and "around midday") had reduced the springtail abundance below the level of before the applications. Only the mechanical treatment showed significant reductions of abundance. The high impact of mechanical applications presented in these findings can be attributed to the habitat destruction caused by the massive soil movement.

The results indicate that, despite the very high earthworm abundance on the selected test area, the use of electrical weeding had no impact on the earthworm population. Neither of the two electrical weed control applications (in the morning and around noon) brought the earthworm abundance significantly below the level of the untreated control variant.

Only the mechanical treatment showed a significant reduction in abundance, which again can be attributed to the destruction of the habitat by soil movement.

The results showed that in the untreated control earthworm abundance was reduced after four weeks, which can be related to a seasonal reduction of activity in the first 20 cm of soil depth at the time of the second sampling at the end of April (the increasing dryness during summer). Comparable to the control, the number of earthworms was reduced in both Electroherb™ treatments four weeks after the application, which brought the abundance to the same level as the untreated control. The greatest effects on earthworms were found in the mechanical treatment (cutter and harrow), which could be in

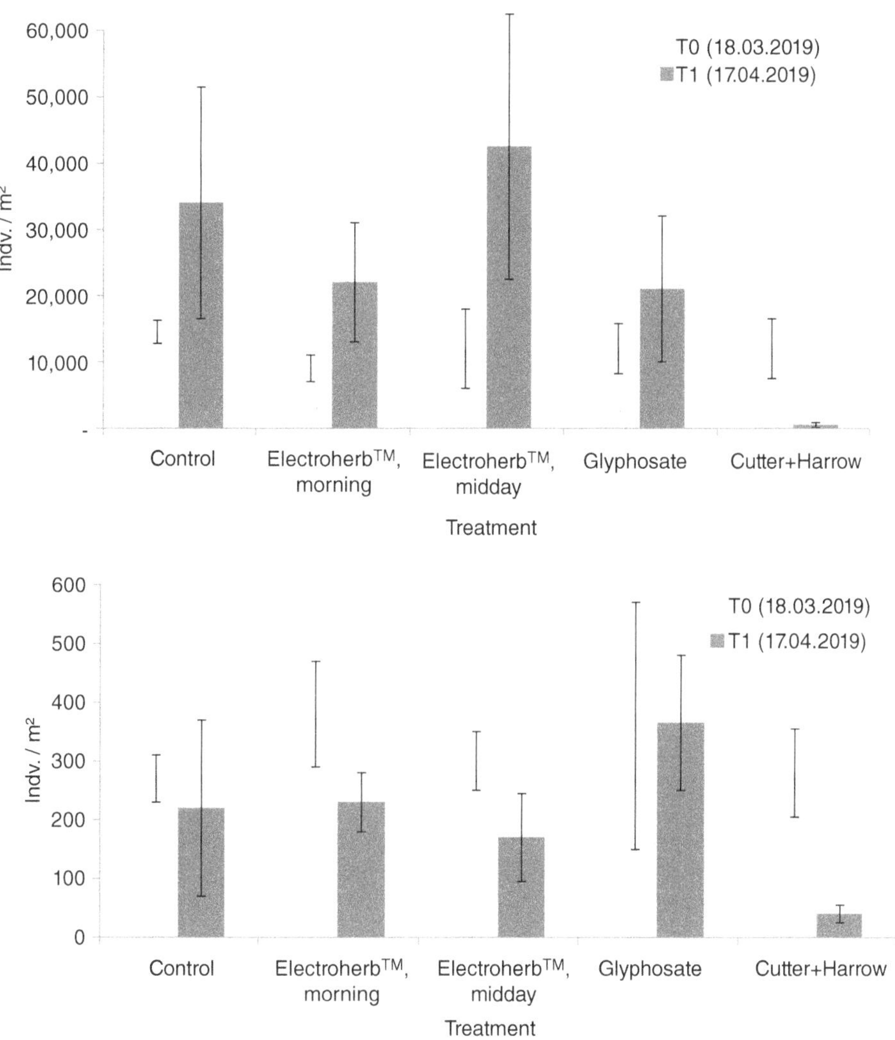

Fig. 7.1. Individuals per area counted for springtails (top) and earthworms (bottom).

connection with the impact from soil movement from the mechanical weed control method. In general, the slight decrease in earthworm abundance (also in the untreated control treatment) could be explained as a seasonal natural variation of activities and abundance in different soil layers.

In conclusion, the study presented here was conducted on an undisturbed grassland area with a very high abundance of organisms per square meter. The results underline the environmental friendliness of the Electroherb™ process, as soil organisms did not experience any lasting damage at the population level. This seems realistic since any damage from electricity in the environment requires a considerable amount of current to pass from the soil into the organisms. However, the electrical transition energy expected to pass from the large soil volume into an adjacent earthworm or smaller organisms with a much smaller cross-section, and sometimes vertically orientated position (consequently with high transition resistances), is very small. The soil, with its large volume and great cross-section, represents an ideal conductor and it is very unlikely that high current densities pass into smaller objects while passing through the soil to the second set of electrodes.

From a theoretical perspective, the impact of Electroherb™ high-voltage current running though the soil with its great cross-section and good conducting properties is very small. In general, the current density is responsible for the heating of a conductor. If the current density is too great in a conductor with too small a cross-section the conductor becomes warm/gets hot. The thermal impact is, furthermore, dependent on the duration of the current flow/applied power. Based on Joule's First Law, we have calculated that the effective power of the Electroherb™ technology, while passing across an arable field with 3 km h^{-1} with 80 kWh applied constant power, results in a maximum temperature change of 1°C in the soil in a worst-case calculation (the calculation of 1°C temperature change relates to 3 kW power running through the first 1 cm of the topsoil layer, which is very unlikely to happen for power propagation in any medium). Accordingly, no effect on the soil matrix, or on soil organisms, can be assumed on cultivated agricultural areas.

The propagation of the voltage in the soil remains close to the system in the case of common homogeneous soils. This is due to a combination of low power density in the soil and short electrode spacing.

Numerous experiments on step voltage evaluation around the system show that theory is reflected in practice. Based on field tests and theoretical consideration by means of computer simulations, a safety distance to the system is defined, which is always based on theoretical worst-case considerations and safety factors, and impedes direct contact with the high-voltage electrodes.

Key finding: the use of electrical weeding had no impact on the earthworm population.

Ecotoxicological Study on an Agricultural Field (by BioChem agrar GmbH, Germany)

This study was conducted to evaluate the environmental effect of Electroherb™ applications on springtails in an agricultural field (mono-maize cultivation; Fig. 7.2). BioChem agrar GmbH, as independent external service laboratory, was commissioned to take samples before, and four weeks and six months after the Electroherb™ application. The study was designed in accordance with the current standards for European Union risk-assessment of agrochemicals. At each sampling date, four soil cores (eight in total) from single treatment plots were sampled and springtail abundance was recorded according to ISO 23611-2 (2006).

In this study, the effect of chemical, mechanical, and electrical vegetation control methods was analyzed. The springtail abundance before the applications in this field averaged 8801±2773 individuals per square meter, which was, compared to the abundance of the first study, relatively high, and assumed to be due to higher soil temperatures in April.

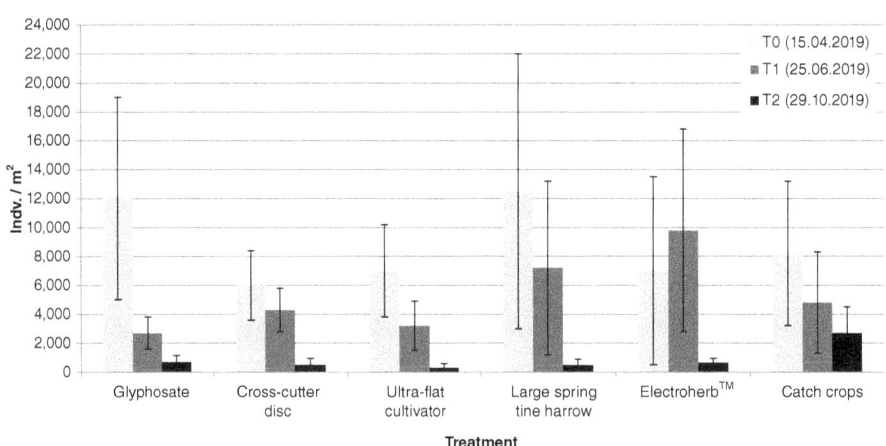

Fig. 7.2 Microarthropod (springtail) units per area.

The results indicated that the Electroherb™ application had no negative effect on the springtail population in the field. The mechanical variants, in relation to their degree of impact on the soil, indicated reduced individual numbers of springtails. This was also the case for the chemical treatment with glyphosate. When compared to the undisturbed catch crop treatment, it is obvious, that four weeks after the treatment the abundance in the Electroherb™ treatment where highest. Similar findings of high springtail abundance after Electroherb™ application were found in several pretests. An increase in dead organic biomass with an increased soil coverage, resulting higher soil moisture conditions, which is beneficial for the development of the springtail population can serve as an explanation for the results in the Electroherb™ treatments of this trial. The results presented here indicate, that after two months, no negative effects on the springtail population were detected for Electroherb™; that cannot be concluded for the mechanical treatments.

Ecotoxicological Study in an Apple Orchard (by CTIFL, France)

Another study was commissioned to investigate the effects on earthworms at the population level of Zasso's XPS and XPO applications in apple orchards, as it is the desired goal to ensure environmental friendliness in this area as well. The average abundance of earthworms in orchards of approximately 15 individuals/m^2 is stated in literature (reported by O. Lysiak in 2015). The sampling of living earthworms was performed by the French fruit and vegetable research organization CTIFL the day after the Electroherb™ applications (XPS: 2.5 km h^{-1}; ~87 kWh ha^{-1}). Applications were performed at normal vegetation control dates in apple orchards during the season, in April, May, and July.

The test was designed to account for typical acute toxic effects on soil organisms, since recovery effects can be excluded by immediate sampling and counting of living organisms the day after application. Fortunately, the earthworm abundance after treatment with Electroherb™ showed that no acute lethal effects on earthworms were found (Fig. 7.3). In conclusion, under realistic field conditions with energy doses that have proven to be effective for vegetation control in orchards, Electroherb™ applications in apple plantations appear to be nontoxic to soil organisms.

The presented study analyzed the acute toxic effects of electrical weeding on earthworm individuals in apple orchards. The results indicated that there were no lethal effects on earthworm individuals one day after the application of common electrical dosages that could be detected. Even though the test area was treated with electrical high-voltage current three times during the growing season, the number of earthworm individuals at all three sampling times was higher or similar to the chemical as well as the untreated control. These results underline

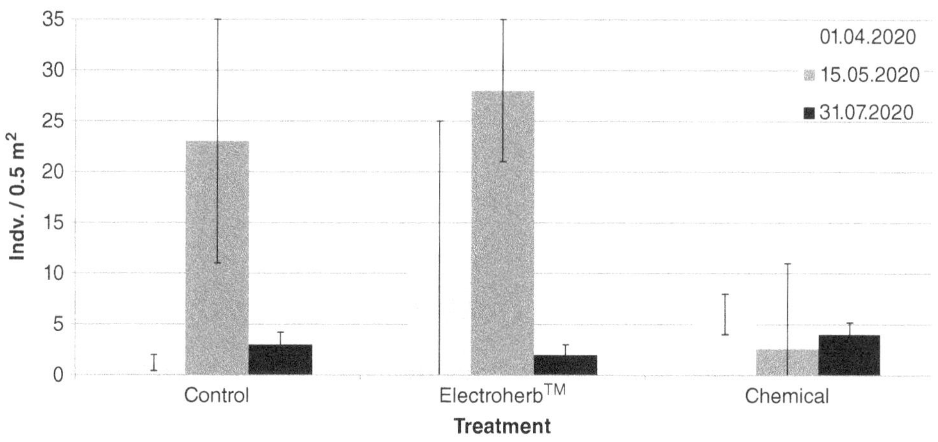

Fig. 7.3. Earthworm units per area.

the conclusions drawn in the previous studies which have focused on term effects on soil organisms after Electroherb™ applications. Thus, as already expressed above, harmful lethal effects from Electroherb™ applications on soil life were not found at this stage of investigation.

Soil Microbial Activity (by UNESP, Brazil)

A study has also been conducted indicating that electrical weeding does not significantly reduce microbial activity in soil. The microbe was *Atividade microbiana* and the study was conducted by the Universidade Estadual Paulista (UNESP) in 2019. The study was carried out at the Faculty of Agricultural and Veterinary Sciences at UNESP, with the objective of assessing whether the application of electrical weeding influences soil microbial activity. The research focused on three primary parameters:

- **Microbial biomass carbon (MBC).** Evaluated through fumigation and chloroform extraction (Fig. 7.4).
- **Basal respiration:** Measured by carbon dioxide (CO_2) emission from soil samples (Fig. 7.5).
- **Dehydrogenase enzyme activity.** Analyzed to determine microbial metabolic activity (Fig. 7.6).

The study compared microbial activity in soil samples before and after electrical weeding treatments.

Methodology

Soil samples were collected from test plots in Jaboticabal, São Paulo, Brazil, before and after electrical weeding treatments. The samples were analyzed using the following methods:

- MBC:
  - Fumigation was performed using chloroform for 24 hours at 25°C.
  - Extraction was done using 0.5M K2SO4.
- Basal respiration:
  - Soil samples were incubated at 30°C for one week.
  - CO_2 emissions were measured to assess microbial activity.
- Dehydrogenase enzyme activity:
  - Soil extracts were incubated at 37°C for 24 hours.
  - After filtration, absorbance was measured at 485 nm using a spectrophotometer.

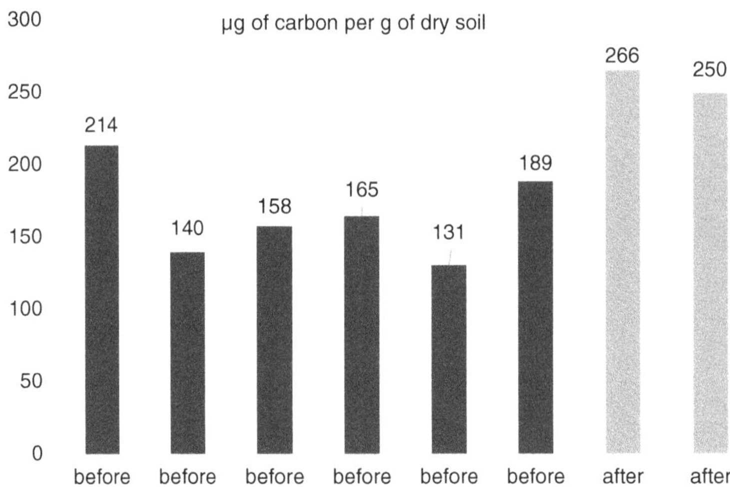

Fig. 7.4. Carbon presence in the soil before and after.

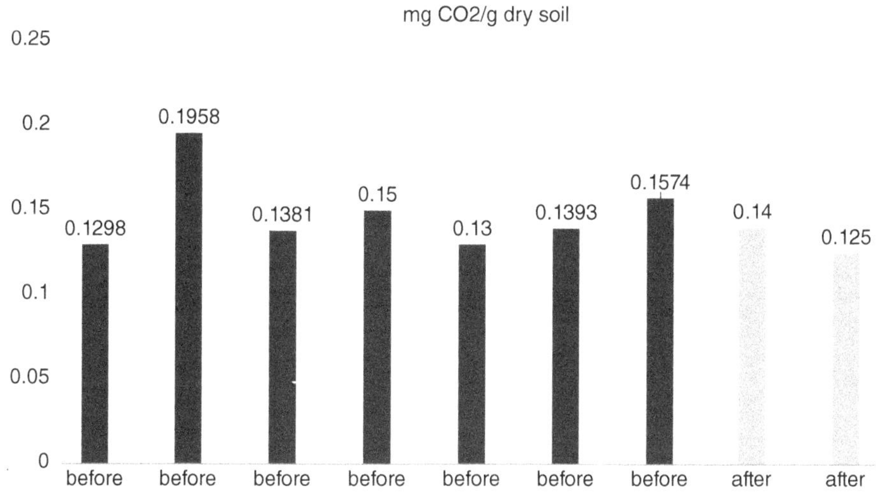

Fig. 7.5. Microbial basal breathing as a measure of microbial activity.

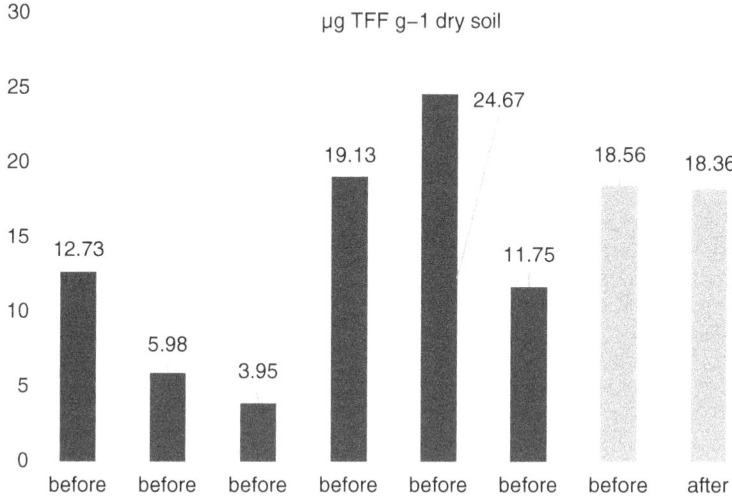

Fig. 7.6. Dehydrogenase as a measure of microbial activity.

The study was structured with two treatments:

- **Before.** Soil samples collected before electrical weeding.
- **After.** Soil samples collected after electrical weeding.
- **MBC.** MBC values before electrical weeding fluctuated between 131.20–214.00 μg C/g dry soil weight. After electrical weeding, the values ranged between 250.00–266.00 μg C/g dry soil weight, suggesting that microbial biomass remained active and even increased after the treatment.
- **Basal respiration.** The results showed variation in CO_2 emission rates among different

samples. The respiration rates before electrical weeding ranged between 0.1298–0.1958 mg CO_2/g dry soil weight, while after electrical weeding, the rates ranged between 0.1250–0.1400 mg CO_2/g dry soil weight. The slight decrease in respiration suggests that electrical weeding did not cause significant microbial suppression.

- **Dehydrogenase activity.** The enzyme activity varied across samples, indicating microbial metabolic function. Before electrical weeding, dehydrogenase activity ranged between 3.95–24.67 µg TFF/g dry soil weight. Posttreatment values were slightly lower but remained within a comparable range (18.36–18.56 µg TFF/g dry soil weight), indicating minimal impact on microbial metabolic processes.

Conclusion

The study results indicate that electrical weeding does not negatively affect soil microbial activity. The microbial biomass carbon levels were stable, and enzymatic activities remained within expected ranges. The findings suggest that electrical weeding is a viable alternative weed control method with minimal ecological impact on soil microbiota. Future research could explore the long-term effects of repeated electrical weeding applications on soil microbial communities.

References

ISO 11268-3 (2014) Earthworms: Sampling: formalin sprouting; manual sorting. International Organization for Standardization, Geneva, Switzerland.

ISO 23611-2 (2006) Springtails: Sampling: soil cores; MacFadyen extractor. International Organization for Standardization, Geneva, Switzerland.

Lysiak, O. (2015) Prélèvements de lombrics, la bonne méthode au bon moment. *WikiAgri*. Available at: https://wikiagri.fr/articles/prelevements-de-lombrics-la-bonne-methode-au-bon-moment/3597/ (accessed November 14, 2025).

8

Safety

On Safety

Electrical weeding involves the use of high-voltage electricity to kill unwanted vegetation by disrupting plant cells and root systems. While it is a chemical-free and environmentally friendly alternative to herbicides, the use of high voltage poses several safety challenges that must be addressed to protect operators, bystanders, and equipment.

One of the primary concerns is the risk of electric shock. Direct contact with high-voltage electrodes or exposed wiring can cause severe injury or even fatal electrocution. Additionally, stray currents in the soil can create step-voltage hazards, meaning individuals standing near the application zone may experience unintended electrical exposure. Ground faults also pose a risk, potentially leading to short circuits or dangerous electrical arcing. To mitigate these risks, proper insulation and shielding are essential, along with the maintenance of adequate safety distances around the application area. Emergency shutdown mechanisms are also a critical safety feature, ensuring that the system can be quickly deactivated in case of an emergency or human proximity to hazardous zones.

Beyond personal safety, the integrity of the equipment itself must also be considered. Overcurrent protection is necessary to prevent electrical surges that could damage the system, leading to operational failures or even fire hazards. Effective thermal management is also essential, as excessive heat generation from prolonged high-voltage application can degrade system components or increase the risk of overheating. Ensuring that electrical weeding systems comply with regulatory safety standards, such as those set by the International Electrotechnical Commission (IEC) and the International Organization for Standardization (ISO), helps establish guidelines for proper grounding, circuit isolation, and overall operational safety.

Environmental and agricultural safety must also be taken into account. If not properly managed, electrical weeding can unintentionally harm nontarget crops or beneficial organisms. Precision application and carefully designed electrodes help prevent electricity from affecting nearby plants. Additionally, controlling electrical conductivity in the soil ensures that essential microbial life and nutrient balance remain undisturbed, preventing potential long-term soil degradation.

Ultimately, ensuring safety in electrical weeding is fundamental to making the technology effective, sustainable, and risk-free. By implementing proper insulation, controlled energy distribution, rigorous operator training, and strict regulatory compliance, the risks associated

DOI: 10.1079/9781836992288.0008

with high-voltage applications can be minimized, making electrical weeding a viable and safe alternative to conventional weed control methods.

The main two risks for human health regard contact with high voltage. This can happen if a bystander directly touches—directly or through a conductive material—the electrodes (Risk A) or the chassis (Risk B) while a fault occurs. Also, safety features can be implemented to diminish the risks.

Risk A

While turned on, someone touches the electrodes (Fig. 8.1). Possible mitigations are: an insulating mat can be added on top of the electrodes in most cases, or a presence sensor can be added in most products to sense if someone is near the electrodes.

Risk B

There is a failure and/or shortcut in the system and the tractor or equipment becomes energized as a whole. Someone simultaneously touches the equipment and the ground, becoming part of the circuit (Fig. 8.2). There are basically two possible mitigations: insulation (double layer preferred) and/or nonconductive materials for high-voltage parts to ensure insulation. Sensors shut the equipment off if it detects insulation failure. Usually, the equipment has its own grounding through which most of the energy flows, diminishing the risk of a higher degree accident.

Both risks are reduced to ZERO if the safety requirement of maintaining 2–3 m from the applicating equipment and any conductive materials touching the electrodes is observed.

Voltage distribution in the soil

When two grounded electrodes are placed in the soil, the voltage distribution between them depends on several key factors, including soil resistivity, electrode placement, and the presence of conductive materials. Understanding how voltage behaves in the soil is essential for ensuring electrical safety, particularly in high-voltage applications such as electrical weeding or

Fig. 8.1. Stick drawing of a "Risk A" type accident.

Fig. 8.2. Stick drawing of a "Risk B" type accident.

grounding systems. The distribution of potential between the electrodes determines step voltages, which can pose a risk to humans or animals in the vicinity of the electrodes.

One of the most significant factors influencing voltage distribution is the distance between the electrodes. If the soil resistivity remains uniform, the voltage drop increases proportionally with distance. This means that if the electrodes are placed at a fixed voltage difference, the voltage gradient between them will follow a linear distribution. For instance, if 1,000 V (1 kV) is applied between two electrodes placed 1 m apart, then at a midpoint of 50 cm, the voltage would measure 500 V. However, this is an idealized case that assumes uniform soil conditions, which is rarely the reality in practical applications. Soil composition and moisture levels often vary, causing nonlinear voltage distribution.

Another critical factor is the presence of conductive materials within the soil. Elements such as buried metal objects, water bodies, or highly conductive minerals can significantly alter the expected voltage distribution. These materials provide alternative pathways for current flow, potentially reducing the effectiveness of grounding and increasing localized step-voltage hazards. In certain cases, conductive objects can even redirect electrical currents, creating unexpected high-voltage zones away from the expected locations. For this reason, a proper site assessment is crucial before installing grounding electrodes or using electrical weed control (EWC) systems in the field.

One of the primary safety concerns in high-voltage soil applications is the step voltage experienced by individuals who might be present in the field. Step voltage refers to the potential difference between two points on the ground where a person might place their feet, leading to an unintended current flow through the body. The magnitude of this step-voltage hazard is influenced by factors such as soil conductivity, electrode spacing, and total applied voltage.

To evaluate human exposure, a 1 kΩ human body resistance model is often used to estimate potential risk. In practice, the voltage that a person experience diminishes rapidly as they move away from the electrodes. If a person is standing directly between two electrodes, they may be exposed to a higher voltage difference than someone standing even a short distance away. However, due to the way voltage dissipates in the soil, the risk drops exponentially with increasing distance (Fig. 8.3).

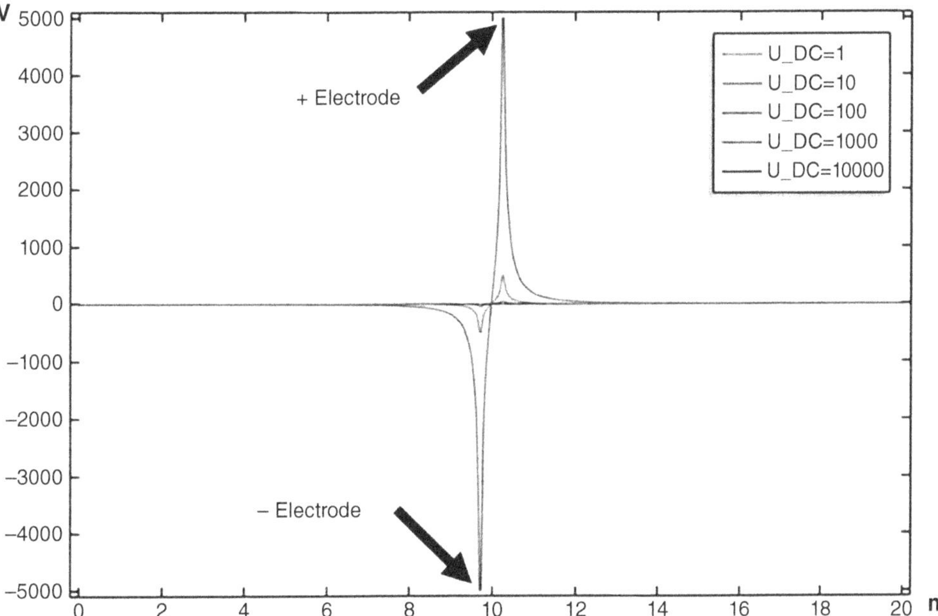

Fig. 8.3. 5 kV application voltage distribution in the soil example.

The behavior of voltage distribution in soil is crucial for grounding applications and electrical weeding technologies. Understanding the relationship between electrode placement, soil resistivity, and conductive materials helps optimize safety measures while maintaining system efficiency. Through proper design, grounding configurations can be adjusted to minimize step-voltage hazards, ensuring both operator safety and optimal system performance. The key takeaway is that while voltage spreads through the soil, its effect on human exposure is highly localized.

The following graphical study was developed by the author and Zasso's engineering team, with special acknowledgements to Mr. Max Jensen.

High-voltage insulation monitoring

The concept of high-voltage insulation monitoring (HVIM) involves the detection of leakage and short-circuit currents on mobile applications with power supply and ground-contacting high-voltage electrodes.

There are multiple ways to supply the necessary electrical energy for an electrical weeding system, each with its own advantages and technical considerations. The choice of power supply depends on factors such as energy demand, mobility, system efficiency, and safety requirements.

One approach is to use a battery-powered system, where a high-capacity battery pack is installed on the carrier vehicle, such as an electric tractor or autonomous vehicle. This option provides a clean, silent, and emission-free power source, making it particularly suitable for sustainable farming applications. To ensure the safe and efficient operation of the batteries, a battery management system (BMS) is required. The BMS monitors charge levels, prevents overcharging or deep discharging, regulates temperature, and ensures balanced energy distribution across the system. The battery system can be modular, allowing for scalability depending on the power requirements of the electrical weeding equipment. However, battery-powered solutions may have limitations in energy storage capacity, requiring recharging or battery swaps for extended operation.

Another option is to use a generator-driven system, which involves a mechanical engine (such as a diesel or gasoline motor) that drives an electric generator. This setup is commonly used in conventional tractors, where a power take-off (PTO) system transfers mechanical energy from the tractor's engine to the generator, converting it into electric power. If a standard tractor engine does not provide sufficient energy, an additional generator can be installed to meet the system's high-voltage requirements. The main advantage of a generator-driven system is its continuous power supply, as long as fuel is available, making it a more reliable solution for long-duration applications. However, this option comes with higher fuel consumption, noise, and emissions, which may be a concern for environmentally focused operations.

From an electrical distribution perspective, the power supply system can be configured as either a TN (grounding "terra"-neutral) grid or an IT (isolated grounding "terra") grid, depending on safety and operational requirements. The TN grid configuration connects the neutral point of the power supply to the ground, ensuring fast fault detection and automatic disconnection in case of an electrical failure. This setup is commonly used in industrial and agricultural applications where grounding reliability is critical. On the other hand, the IT grid system is designed with an ungrounded or high-impedance grounding, allowing the system to continue operating even in the presence of a single insulation fault. This enhanced fault tolerance makes the IT grid a preferable choice for high-voltage applications in harsh or isolated environments where immediate shutdowns due to minor faults are undesirable.

In summary, the choice of electrical energy supply for an electrical weeding system depends on factors such as availability, sustainability, energy demand, and safety considerations. Battery systems offer a silent and eco-friendly alternative, whereas generator-driven solutions provide a continuous and robust power source. Additionally, selecting TN or IT grid configurations ensures that the system meets the safety and reliability requirements necessary for efficient and safe EWC operations.

For design and safety reasons, electric cars are always developed as IT grids. Insulation monitoring devices (IMDs) ensure that the grid is always isolated from the chassis. Other applications for IT grids are supply-critical facilities such as hospitals, where a single-pole earth fault does not lead to grid failure. With TN grids, as they are usually used in the supply grid, a residual current can be detected via residual current device (RCD). An RCD type A, which represents a 30-mA residual current protective device, is usually used for this purpose. In applications where a direct current (DC) component in the residual current is generally also to be expected (e.g. charging of an electric car), an RCD type B (all-current sensitive) can be used or an residual current monitoring unit (RCMU) can be connected upstream.

Problem formulation

Due to capacitive coupling, e.g. in the area of the high-voltage transformer or due to single-pole short circuits, states can be achieved in which the chassis of the carrier vehicle is raised to impermissible voltage levels. Because the vehicle is not grounded (tires are only approximately insulating), it can initially be regarded as isolated from the ground. Single-pole short circuits between the high-voltage side and the chassis are particularly critical. These can occur, for example, due to insulation failure in the area of the transformer, the high-voltage cables, the electrodes, or any connections and connectors in the high-voltage circuit. The reason for this can be incorrect design of the insulation concept, pollution, or the ageing mechanisms of the insulation material. Electrical treeing due to material impurities (air inclusions, solid inclusions) can lead to material fatigue in the short, medium, and long term and cause an increase in leakage currents or start breakdown mechanisms. If liquid insulating materials (e.g. oil-insulated transformers) are used, fiber-bridge breakthroughs or increased moisture or pollution can lead to failure of the insulation property in addition to overloading.

With the application presented Fig. 8.4, there is an additional risk that the ground vegetation will touch both an electrode and the chassis of the carrier vehicle and thus directly cause a single-pole short circuit. The voltage level to which the chassis is increased in this case depends primarily on the electrode voltage, the conductivity of the plant, and the transition resistance from the applicator to the plant and from the plant to the chassis of the carrier vehicle.

Using the examples of the TN-S grid and IT grid, initially a no-fault current can be detected with the single-pole short circuit. Both the normally used RCD technology and the IMD technology, which detect residual currents on the low-voltage side, are not active with high-voltage-side short circuits.

Generation and detection of fault currents

To generate a detectable fault current and to reduce the chassis voltage in the event of an insulation fault, the vehicle has an earthing element in combination with insulation monitoring (Fig. 8.5). The insulation monitoring consists of

Fig. 8.4. Representation of the single-pole short circuit, e.g. direct contact of the electrode with the chassis.

Fig. 8.5. Representation of the current distribution in effective current flow and residual current flow to be detected.

a low impedance, all-current sensitive current measurement with corresponding evaluation electronics (RCMU). An RCD type B can be used for this purpose, whereby only the earth conductor is detected.

The fault current flows from one electrode, through the ground, the earthing, via the chassis to the location of the short circuit, and to the second electrode. For systems with floating potential, the chassis voltage can be reduced depending on the grounding impedance.

The size of the fault current depends on the voltage applied to the electrodes, the type of short circuit, the impedance distribution in the ground, and the transition impedances of the electrodes to the plant (or ground) and from the ground to the vehicle.

The fault current $\underline{I}_{fault}$ can be determined as a function of the total current $\underline{I}_{total}$ (Eqn 8.1):

$$\underline{I}_{fault} = \underline{I}_{total} * \frac{\underline{Z}_1}{\underline{Z}_1 + \underline{Z}_2}$$

Equation 8.1. Fault current as a function of total current.

The fault current can be determined as a function of the effective current $\underline{I}_{eff}$ (Eqn 8.2):

$$\underline{I}_{fault} = \underline{I}_{eff} * \frac{\underline{Z}_1}{\underline{Z}_2}$$

Equation 8.2. Fault current as a function of effective current.

Since the residual current depends on the total current, a minimum total current can be defined to generate a fault current in the faulty system. This must be reached during operation so that fault detection can become active in the event of a fault. In operation, this can be achieved by defining a shutdown condition in which the total current (or the output power) is averaged over a period of a few seconds (e.g. >1 s). If the value falls below the limit value, the system is switched off and the operator must act to put the system back into operation. If, for design reasons, it can be ensured that no lethal body current flows when touched, the switch-off condition can be dispensed with.

Isolation of the application space

The possibility of short-circuiting due to plants touching both an electrode and the chassis can be counteracted by making the application space completely isolated from the chassis potential (Fig. 8.6).

To ensure electrical safety and system integrity, the application space is insulated from the machine's frame using high-voltage-resistant plastic materials. These insulating materials are carefully selected to withstand high electrical stresses while preventing leakage currents and unintended electrical discharge.

The plastic assembly is specifically designed to eliminate any direct conductive connection between the application area—where high voltage is applied—and the rest of the machine. By doing so, the system effectively prevents voltage transfer to unintended parts, reducing risks

Fig. 8.6. Illustration of the definition of the application space.

associated with electrical faults. These are the key insulation features:

- *Air and creepage distances* are meticulously maintained to prevent unintended discharge or breakdown of insulation.
- *All fastenings use blind holes*, ensuring that no exposed conductive elements penetrate the insulation, thereby enhancing safety.
- In the event of *wiring damage* due to vibrations, mechanical stress, or plant interference, the fault remains *contained within the high-voltage section*, preventing accidental electrical exposure to other machine components.

To further mitigate electrical hazards and prevent impermissible voltage potential from affecting the machine frame, the entire front frame can be constructed using nonconductive materials, such as glass-reinforced plastic (GRP). This ensures that any potential insulation failure does not result in hazardous voltages on the machine's support structure.

For autonomous vehicles or specialized applications, the entire chassis can be manufactured from nonconductive materials, ensuring that all machine components remain electrically isolated. This approach enhances operator safety, minimizes electrical interference, and ensures compliance with high-voltage insulation standards in precision agricultural applications.

Insulation of the vehicle chassis

High ground cover, which touches the chassis of the carrier vehicle, reduces the probability that an insulation fault will be detected within a short time. Due to the contact, a fault circuit that flows parallel to the installed grounding is closed and cannot be detected (Fig. 8.7).

To reduce the proportion of undetectable fault currents, the underside of the chassis and all other attachments are insulated accordingly so that the plants cannot come into contact with the chassis (Fig. 8.8). This can be designed as pure contact insulation (no high-voltage insulation), as a high voltage of the chassis (already <1 kV) is detected by HVIM. If this fails, grounding via ground vegetation can reduce the probability of an applied high voltage.

Energy distribution and fault localization

Using modular energy distribution in high-voltage systems improves both safety and biological performance by splitting the available output across several independent modules. If accidental contact occurs, the system architecture ensures that only the energy of the single involved module can couple into the unintended path rather than the full system output, which sharply reduces the likelihood of injury or fatal electrocution. Each module is intentionally power-limited so that unintended paths cannot sustain excessive current. Depending on the design, modules may present one or multiple poles on their input side; these further shape how energy is apportioned and lower the probability that a fault current reaches lethal magnitude.

The same modularity also enhances field efficacy. Because energy is metered per module,

Fig. 8.7. Splitting of the fault current into detectable and nondetectable components.

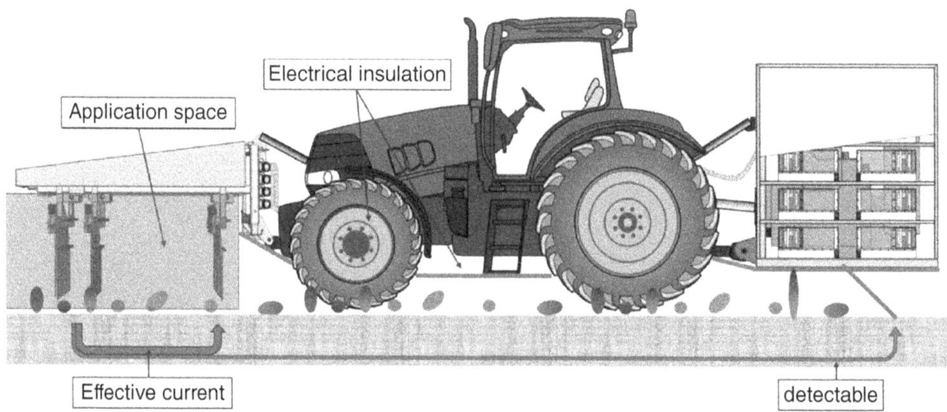

Fig. 8.8. Electrical insulation of the vehicle and all attachments.

power can be distributed more evenly across the application area, promoting consistent plant desiccation. To avoid high inrush currents when the system is turned on, the modules start in a short, time-staggered sequence that gently ramps the total load and reduces component stress while extending service life. This delay should remain very brief so treatment continuity is not compromised during motion; in practice all modules should reach full operation in under one second to maintain seamless coverage without sacrificing operational efficiency. In this way, the modular approach balances safety, efficiency, and biological effectiveness and is therefore a core design feature for modern high-voltage agricultural equipment.

If an insulation fault occurs in an inverter's housing, the system can localize the fault by using inverters whose enclosures are isolated from the chassis. Residual current monitoring (RCM) then identifies the faulty unit. It is important to distinguish between a fault on the low-voltage side and one on the high-voltage side; RCM is used to detect insulation faults on the low-voltage side. In such a design, two sensors—configured with universal current sensitivity—monitor leakage so the system can flag and safely shut it down.

Design of the earthing element

In electrical weeding applications, proper earthing is essential to ensure safe and efficient operation. The earthing element establishes a connection between the chassis of the carrier vehicle and the ground (earth potential). This connection provides a return path for electrical current, preventing unintended voltage buildup and reducing electrical hazards. A well-designed earthing system must exhibit low resistance and low earthing resistance to facilitate effective energy dissipation while also minimizing its impact on the maneuverability of the vehicle.

A chain is a widely used earthing method due to its high flexibility and minimal impact on maneuverability. The loose structure of the chain ensures constant contact with the ground, even on uneven surfaces. However, the individual chain links introduce contact resistance, which can reduce the efficiency of the earthing connection.

To overcome this limitation, a cable (e.g. copper flat strips or braided conductors) can be threaded and screwed through the individual chain links, ensuring better conductivity and lower overall resistance, or used instead of the chain. The addition of extra weights may further enhance the ground contact, improving the earthing resistance.

Unlike chains, cutting knives or plates offer a more direct electrical connection with the ground, resulting in a lower earthing resistance. These elements are pressed into the soil, creating a firm and conductive contact. The main advantage of using cutting knives is their superior conductivity compared to chains or mats, making them highly effective in conditions where low earthing resistance is crucial. However, their disadvantage is that they can limit the vehicle's maneuverability, as they must maintain continuous soil penetration to function effectively. This restriction makes them less suitable for fast-moving or highly mobile electrical weeding systems.

Conductive mats provide a broad area of contact with the ground, which significantly improves the stability of the earthing system. The larger the mat's surface area, the lower the earthing resistance, allowing for better energy dissipation. One key advantage of mats is that they can be weighted down to increase ground contact. Additional weights of various shapes and sizes can be attached to the mat to improve soil penetration and enhance its conductivity. However, similar to cutting knives, heavier mats may limit the vehicle's mobility, making them more suitable for stationary or slow-moving applications.

A cable with a metal tip or mat attachment provides another effective means of grounding the electrical weeding system. The metal component at the tip ensures direct soil contact, reducing the contact resistance and enhancing electrical conductivity. This design offers moderate flexibility while still maintaining adequate ground penetration.

The choice of an earthing method depends on several factors, including the terrain, the required earthing resistance, and the mobility of the vehicle. While chains offer maximum flexibility, cutting knives provide superior conductivity, and conductive mats cover a larger area for efficient grounding. A well-designed system may incorporate a combination of these methods to optimize both safety and operational efficiency in EWC applications.

Monitoring of the earthing element

A complete loss of the earthing elements (e.g. due to tearing off or dismantling) or faulty commissioning (fixing of the earthing elements for transport) may result in the chassis not being connected to the ground. For this reason, it must be ensured that the earthing elements are in their intended target positions during operation.

Loop monitoring can ensure that mechanical loss or damage to the grounding elements is detected. A quiescent current (typically a few milliamperes) is coupled into a permanently wired, low-impedance circuit and measured. If the circuit is interrupted at one point or closed, this can be detected. Corresponding electronics (loop-monitoring relays) can be purchased as a standard component and can be integrated into the concept presented here. If a fault is detected, the entire system can be brought into a safe state, e.g. via a programmable logic controller (PLC) or safety PLC.

Basically, the concept can be integrated at any position. A cable (or an electrical conductor/return conductor) is laid parallel to the earthing element, and is electrically insulated from the earthing element. Only at the point where the earthing element touches the ground (or as close as possible to the point) is the cable electrically connected to the earthing element. This ensures that a partial loss of the earthing element is also detected. In the event of a fault, the closed circuit is interrupted, and the electronics report a fault. The PLC can then switch off the high voltage at the electrodes and bring the system into a safe state. There is only one electrical connection to the chassis.

The earthing elements are connected in series for this purpose. This ensures that any loss of any grounding element interrupts the ground loop and fault detection takes place. In principle, parallel connection is also possible.

If no loss of the earthing element is to be expected, assembly or disassembly monitoring can be carried out in the same way. The connection between earthing element and return conductor (e.g. cable) can be made close to the frame.

If the entire system is to be transported, it is possible to simplify handling by using fastening points. These fixing points are used to fix the earthing elements for transport. If the entire system is prepared for operation, the commissioning engineer must manually loosen the fastening elements (or the system automatically loosens them).

In order to monitor this process, a second type of loop monitoring can be used. The fixing points must be electrically isolated from the chassis for this purpose. When used, the fixing points are electrically connected to the earthing elements. The fixing points are at a sufficient distance from the suspensions of the earthing elements to prevent an electrical connection between them (if the earthing elements are not fixed). The fixing points can be installed at any position on the frame. The fixing points are connected in parallel to ensure that all fixing points have been properly loosened.

The monitoring circuits work independently of each other and can be used individually. It must be ensured that there is a connection to the chassis in each case.

The earthing element is manually connected to a pull-wire switch or lever switch for transport via a spring element (optional). This ensures a defined installation of the earthing element. A pull-wire switch can also be used in combination with a lever. The switch is attached to the frame/chassis. In principle, the positioning of the switch is arbitrary and depends on the length of the earthing element. The switch has an automatic reset. A mechanical spacer can be added as protection between the frame and the earthing element.

If the entire system is to be prepared for operation after transport, the earthing element must be disconnected from the spring element or switch manually or automatically. If this does not happen or happens incorrectly, this is detected by the switch and forwarded to the electronics (e.g. PLC). This can then prevent the entire system from starting up and provide the operator with an error message and/or a note for action. The only way to attach the earthing element is to fix it via the pull-wire switch.

Risk of direct human contact suppressor

Danger zone barrier

The "safety distance" is defined by the condition that the electrodes must not be touchable during operation, i.e. that nobody must approach the high-voltage range. Furthermore, the potential distribution in the ground has an influence on the step voltage. Far away from the application area, however, this is so small that no further monitoring or shut-off is necessary. Danger Zone 2 results from the consideration that there is intuitively nobody in front of and behind the front frame in the moving system since the danger from the moving carrier vehicle is obvious (Schematic 8.1). This area is therefore monitored by the driver and not additionally blocked off. Since the density of the equipotential lines is highest in the ground at the side of the front frame in most applications and direct contact with the electrodes cannot be ruled out, Danger Zone 1 can be defined accordingly. The Danger Areas 1–3 apply to every application area of a machine. There is no need to install an emergency stop switch near an application area, as this would be located within Danger Zone 1 or 2. This area should also not be entered in case of danger.

The barrier for Danger Zone 1 can only be added if there are no tall plants in this zone (e.g. fruit orchards, vineyards). The barrier consists at least one linkage which can be pushed into the frame or can be collapsible for transport purposes. In addition, redundantly designed sensors can be used to record the status (pushed in or out) and report it, e.g. to the safety PLC. At least level C in the safety assessment, as defined in ISO 13849-1, can be reached. The frame can be

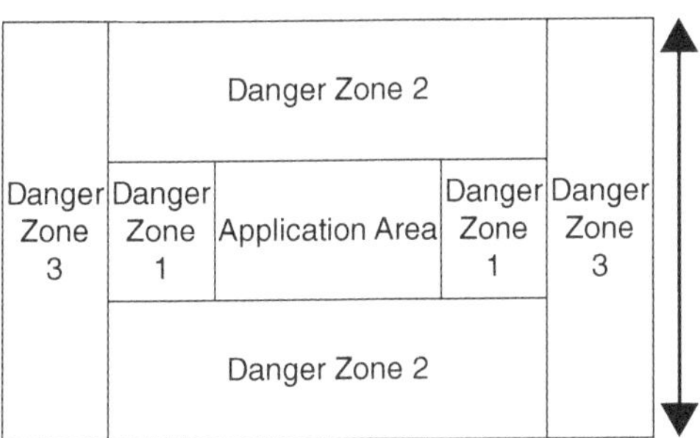

Schematic 8.1. Definition of danger zones.

covered with at least one mat. Another way to close off Danger Area 1 is to use at least one rope in combination with a linkage and at least one pull-wire switch.

Ensuring the plant or ground contact of the electrodes during operation is another hurdle that makes it difficult to touch the electrodes during operation. In order to guarantee the safety precautions of the attachment independent of the carrier vehicle, several combinable solutions are presented. The high voltage can only be switched on when the sensor(s) confirm the target position of the electrodes. If the target position of the electrodes is left on during operation, the high voltage is switched off.

Top link position monitoring

This approach has a connection to both the top link and the front frame. A sensor is switched on as soon as the front frame is in the target position by means of a frame which adjusts itself depending on the angle between the two elements. A torsion spring presses the mechanics permanently in the direction of the tractor against a stop which is fixed in the longitudinal groove. A coded radio-frequency identification (RFID) sensor, with level D release, is arranged at the outermost point of the mechanics. A second arm is also rotatably mounted on the mechanism. The actuator is located at the outermost point of this arm. The arm is attached to the top link of the tractor with a rubber band so that it moves when the linkage moves from the tractor. Alternatively, other switches can be used.

Top link position monitoring has at least one connection to the top link, as well as to the frame of the application. Optional, it has at least one spring and a moving mechanism that adjusts depending on the angle between the top link and the frame. Top link position monitoring also has at least one (micro-) switch or distance sensor which detects the target position. In principle, this arrangement can be attached to any position of the overall system that is suitable for it.

Electrode pressure sensor

A construction can be used to detect the electrode status, which makes it possible to detect the position of a moving electrode. The electrode is mounted on a suspension which allows the electrode to move. The position of the electrode depends on whether the electrode is on the ground or in the air. To detect this position, a high-voltage insulating rod is used, which changes the status of a button or pressure sensor. This status change is detected by an electronic evaluation system. Only when the electrode is in contact with the substrate can the application start operating. If the ground contact is lost during operation, this automatically switches off the application.

Open-circuit detection

The circuit that determines the biological effect of the application is defined by a plus electrode and a minus electrode. The circuit is not closed until the electrodes are in contact with the substrate. This change of state can be used to detect whether the electrodes are in the target position, i.e. have ground contact, by means of current measurement. If there is voltage at the electrodes and the measured current is approximately 0 A, it can be assumed that the system is not ready for the application. This can have two causes: the electrodes are not in the target position or the ground is too high resistance for a current flow to occur. In both cases, an application is not effective. In principle, the full operating voltage is not required to generate a test current. The necessary test voltage can be derived from the fact that the ground conductivity must fall below a certain value for the application. In conclusion, this means that if the test voltage is switched on, a limit value for the current flow must be exceeded in order to start the application. A further determination of the test voltage can represent safety-related restrictions. If the electrodes are not in the target position, there must be no dangerous test voltage at the electrodes. For this reason, the test voltage can be a high-frequency alternating voltage or a pulsed voltage. A power evaluation during operation can be used to detect a possible departure from the target position. If a limit value is not reached during application, operation is automatically exited, and the user is informed that either the substrate has too high impedance or that the target position of the electrodes has been left.

Ultrasonic and radar sensor

To measure the frame height, at least one ultrasonic or radar sensor is used on the frame, which measures the distance between the ground and the frame. This is attached to the frame in such a way that the measurement signal leaves the sensor constantly in the direction of the substrate. The angle between measuring signal and ground is 90°. In case of unevenness, this angle can deviate by several degrees. Basically, several ultrasonic sensors can be used on each frame to achieve redundancy. Basically, the sensors can be mounted at any height of the frame and detect objects that are within the measuring funnel at a mostly adjustable value to the sensor.

If the ground is overgrown, a distinction can be made between different height levels. In most applications, the height from the ground is usually decisive for the safety assessment.

Human obstacle presence sensors

In order to monitor the danger zone created using high voltage around the entire system, camera systems can be used that can detect the entry of humans (or animals). The danger arises both from potential propagation in the ground and from direct contact with the electrodes. In order to detect the presence of people in the danger zone, the camera system is mounted on a frame or on the carrier vehicle. As soon as a person enters the danger zone, this automatically leads to the shutdown of the operation. Operation can only be started (manually) if no person is present in the danger zone.

Height-adjustable wheels

If the frame, which contains the electrodes, is not held by a lift as part of a drive vehicle, but has its own wheels, these can be implemented in a height-adjustable way. For transport purposes, the electrodes can be lifted so that they do not touch the ground during transport. In addition, the height of the electrodes can be adjusted depending on the height and the nature of the vegetation. Height-adjustable wheels can be implemented at any position of the system. This setting influences the contact pressure on the plant, which, depending on the application, represents one optimization parameter (amongst many others).

Risk of shutdown malfunction suppressor

On the one hand, the safety circuit should be designed to stop the operation of the system if an unsafe condition of the entire system is reached and, on the other hand, to prevent the entire system from starting up if in an unsafe condition. The status of the entire system should be detected by sensors and confirmed by the operator via the status of the emergency stop switches. A central main safety control unit should be used to record all necessary sensor outputs, which releases the contactor, an electrical component that turns the equipment on or off. Due to the size and composition of the overall system, sub safety units are used to reduce the cabling effort by means of sub-distributions. All connections can be exchanged at will using identical plug/socket combinations.

All sensors and emergency push-off buttons should be connected in series. To create redundant measurement signal acquisition, identical sensors of different types or technologies can be connected in series.

Alternating current vs direct current safety-related issues

Why alternating current systems have higher voltage peaks and why it matters

For the same power delivery, and therefore the same efficacy of application, synodal AC systems require much higher voltages. Higher voltages are less safe for humans due to several factors related to how the human body reacts to electrical currents (Fig. 8.9), the two main points being:

- **Increased tissue damage.** When exposed to higher voltages, more current can flow through the body, potentially causing greater tissue damage. This can lead to burns, muscle contractions, and damage to internal organs.
- **Arcing and flashover.** Higher voltages are more likely to cause electrical arcs or flashovers, which can result in sudden, intense releases of energy. These phenomena can cause severe burns and blast injuries, even at some distance from the source of the electrical energy. And the higher the voltage, the higher the distance you must maintain from the source.

For the same performance, alternating current (AC) needs much higher voltage and current peaks. This effect is greatly increased if the power control is not done through voltage adjustment. The light blue line shows the current when a piece of equipment is set for a high impedance but used at low impedance. The actual resistive system that includes a human is slightly more complex, but the concept remains the same.

AC is more commonly associated to the "can't let go" effect. In cases where the conductor is grabbed by hand, the hand contracts its muscles and the person cannot let go, leading to a very dangerous accident.

To reach the ventricular fibrillation threshold, the current through the heart system must be at least 100 mA. According to Fish and Geddes (2009) current peaks are key to determining the damage potential of the accident, as shown in Table 8.1.

The combination of pulse density modulation (PDM) and pulse width modulation (PWM), implemented in some of the products can be used to create a "nonapplication" period every few milliseconds, which would give the opportunity for the subject of an accident to "let go" of the conductor. Moreover, this PDM system greatly reduces the size of the "arcing" potential since it deionizes the air several times per second.

Due to higher voltage peaks, the safety distance is increased when AC is used, since the higher voltage (together with other effects) can break the dielectric of air and other insulators with greater ease when compared with DC.

Controlling constant power through voltage ensures not only a series of efficacy and efficiency advantages, but also the minimum application of voltage and current peaks, as per Fig. 8.9. Basically, this is due to not having to set up a higher-than-needed voltage at any specific point in time to ensure the power needed to perform electrical weeding.

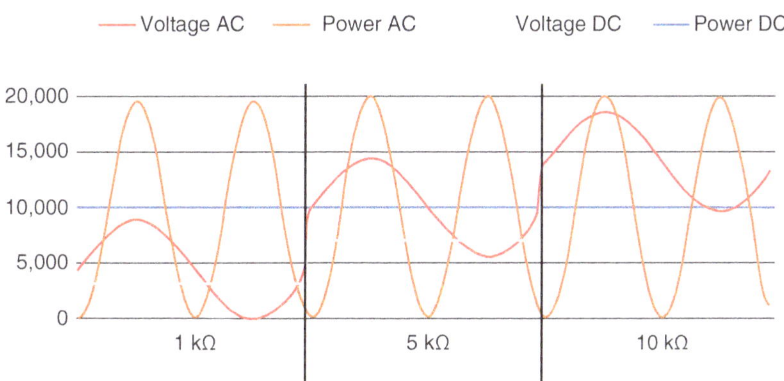

Fig. 8.9. Behavior of voltage between alternating current (AC) and direct current (DC) at constant power for 1 kΩ, 5 kΩ, and 10 kΩ.

Skin effect and its safety implications

There have been claims suggesting that high-frequency AC is "safer" than DC, particularly due to the skin effect—a phenomenon where electric current tends to concentrate near the outer surface of a conductor rather than flowing uniformly throughout its cross-section. The reality is far more complex and nuanced, and in some cases, high-frequency AC exposure may pose even greater risks, particularly in regard to cardiac events.

In a typical conductor, such as a copper wire, the skin effect causes higher current density near the surface, reducing the current penetration into the core of the material (Fig. 8.10). This is a well-documented phenomenon in electrical engineering, where high-frequency AC reduces the effective cross-sectional area of conduction, leading to increased resistance at higher frequencies.

Table 8.1. Current peaks and accident damage potential. (Fish and Geddes, 2009.)

Current peak	Accident damage potential
1 mA	Barely perceptible
16 mA	Maximum current an average man can grasp and "let go"
20 mA	Paralysis of respiratory muscles
100 mA	Ventricular fibrillation threshold
2 A	Cardiac standstill and internal organ damage
15/20 A	Common fuse breaker opens circuit

However, applying this concept to biological tissue introduces critical complications.

The human body is not a uniform conductor like a metal wire. Different tissues have varying electrical conductivities, and the distribution of current is far from uniform. Blood is one of the best conductors in the body, as it contains ions that facilitate electrical flow. Conversely, bone, dry skin, and fat are much more resistive, causing current to flow preferentially through low-resistance pathways, namely, blood vessels and muscle tissue.

One of the most vulnerable muscles is the heart, which relies on precise electrical signaling to maintain proper rhythm. As current concentrates around blood vessels, it increases the likelihood of interfering with cardiac muscle function, potentially triggering arrhythmias, or even cardiac arrest. This contradicts the assumption that high-frequency AC is inherently safer, as it actually increases the likelihood of heart-related complications rather than reducing them.

Understanding the actual impact of high-frequency AC on the human body is essential for electrical safety protocols. While DC tends to create a more continuous flow of current, high-frequency AC can rapidly alternate, potentially causing more severe muscle contractions and cardiac disruptions. Additionally, because blood vessels serve as efficient conductors, localized heating and increased current density around them could further exacerbate tissue damage.

Fig. 8.10. Current flow in low- and high-frequency currents.

The interplay between frequency, current intensity, tissue conductivity, and exposure duration determines how hazardous electrical exposure truly is. The belief that high-frequency AC is always safer due to the skin effect is misleading, as its tendency to affect the periphery of blood vessels makes it particularly dangerous for muscle-rich areas, especially the heart. Thus, electrical safety should not rely on simplified assumptions about frequency effects but should instead prioritize thorough research and safety measures, ensuring protection against both AC and DC exposure risks in high-voltage environments.

Reference

Fish, R.M. and Geddes, L.A. (2009) Conduction of electrical current to and through the human body: A review. *Eplasty* 9; e44.

9

Publications on the Efficacy of Electrical Weeding

Inter-row Electrical Weeding in Organic Cotton

The following is based on "Insights on inter-row electrical weeding as a non-chemical weed management tool in organic cotton." by Ryan Hamberg, from Texas A&M University (Hamberg, 2025).

Organic production systems traditionally rely on inter-row cultivation as a primary tool for weed management. While effective in suppressing weed growth, mechanical cultivation introduces significant trade-offs, such as repeated soil disturbance, erosion, degradation of soil structure, and loss of organic matter. These side effects run counter to the core principles of organic agriculture, which prioritize soil health and sustainability. Thus, the pressing need arises for weed control technologies that are both effective and soil-conserving.

The Texas A&M University Research Farm is now home to a prototype inter-row electric weeding system specifically designed to minimize soil disturbance while effectively controlling weeds. Developed in collaboration with Zasso and supported by AgriLife Research and Cotton Incorporated, this small-plot research prototype integrates a high-frequency power unit mounted at the rear of a tractor, with applicator electrodes mounted on the front loader.

The goal of the Texas A&M research program was to assess the performance of the electric weeding system compared to traditional inter-row cultivation under early post-emergence conditions (cotton at the 4-leaf stage). Several configurations were tested based on:

- **Travel speed.** 1.3, 3.2, and 5.6 km h^{-1}.
- **Number of passes.** single vs double (immediate or delayed by 3 days).
- **Comparators.** traditional inter-row cultivation and nontreated control.

The fields under evaluation were infested with a representative mix of common broadleaf and grass species, including Palmer amaranth, ivy-leaf morning glory, hophornbeam copperleaf, Johnsongrass, and Texas panicum. Weed heights ranged between 5–10 cm.

The system demonstrated excellent efficacy on broadleaf species, achieving over 95% control across nearly all treatments at 3 days (Fig. 9.1) and 14 days (Fig. 9.2) after treatment (DAT). Performance was comparable to mechanical cultivation in most cases, particularly at lower speeds and with two-pass applications. The only scenario where performance dropped noticeably was at 5.6 km h^{-1} with a single-pass application, suggesting that higher speeds may compromise efficacy on some weed species.

Grass weed control was more variable and sensitive to application parameters. At 1.3 and 3.2 km h^{-1}, two-pass electric weeding treatments performed slightly better (by approximately 3%) than a single pass of cultivation. However, single-pass electric weeding at 5.6 km h^{-1} saw a

DOI: 10.1079/9781836992288.0009

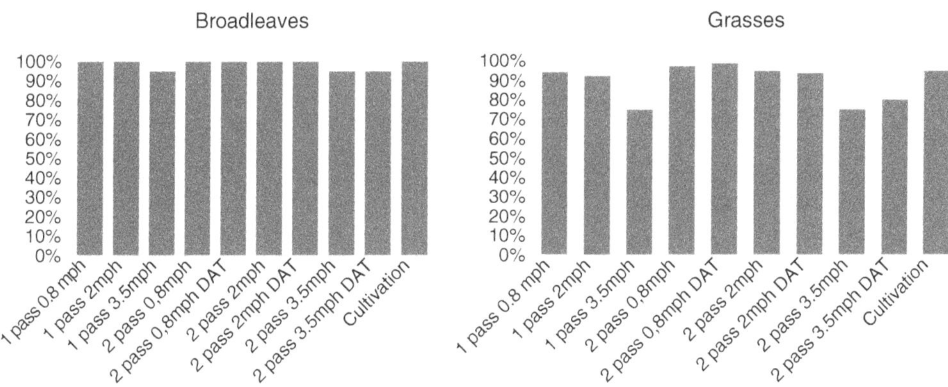

Fig. 9.1. In-row electrical weeding control 3 days after treatment (visual control %).

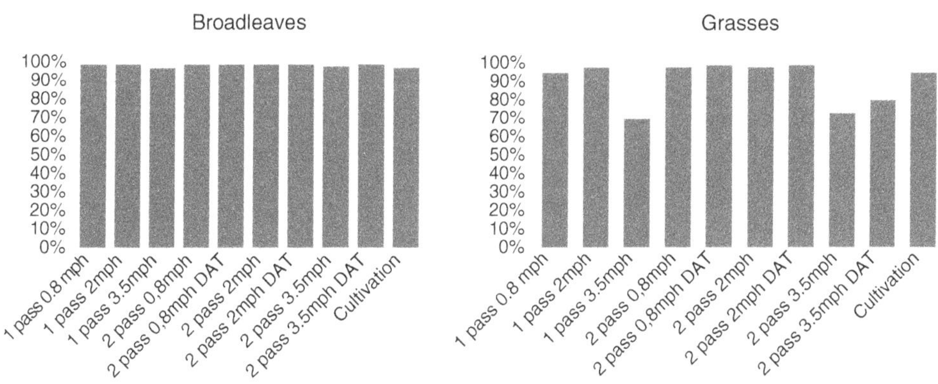

Fig. 9.2. In-row electrical weeding control 14 days after treatment (visual control %).

marked reduction in control (70%), with even two-pass applications at this speed yielding only 74–81% control. These findings are consistent with prior research indicating that grass species, due to their root system morphology, are more tolerant to thermal and electrical damage.

Cotton injury assessments at 3 and 14 DAT showed minimal impact, never exceeding 1% of the plot area. All observed injury was attributable to physical contact between the applicator and the crop, rather than unintended electrical transfer via the root system. This suggests a favorable safety margin for precise applications.

The early results from the Texas A&M study support the hypothesis that inter-row electrical weeding can serve as an effective nonchemical weed control solution in organic cotton production. Its ability to rival mechanical cultivation in broadleaf suppression while preserving soil health positions it as a transformative tool for organic and regenerative systems.

Management of Italian Ryegrass (*Lolium multiflorum*) in Hazelnut Production

The following is based on "Nonchemical approaches to managing Italian ryegrass in hazelnut orchards" by Marcelo L. Moretti, from the Department of Horticulture of the Oregon State University, with the support of the United States Department of Agriculture (USDA; Moretti, 2022).

Italian ryegrass (*Lolium multiflorum*) is recognized as the most problematic weed in hazelnut production due to its aggressive growth, adaptability, and resistance to multiple herbicide modes of action. This invasive species is primarily managed using chemical herbicides, applied both pre-emergence and post-emergence. However, concerns regarding herbicide resistance necessitate the exploration of alternative, nonchemical control strategies.

Current management practices for controlling Italian ryegrass in hazelnut orchards rely on chemical and mechanical approaches, with an increasing emphasis on community-level strategies due to the challenge of herbicide resistance. Chemical control remains the primary method, with glufosinate being one of the most effective herbicides for managing ryegrass. Pre-emergence applications help prevent germination, while post-emergence treatments target established plants. However, resistance to glyphosate, clethodim, and paraquat has been reported, reducing the effectiveness of traditional chemical approaches.

Mechanical control methods such as tillage are generally avoided in nut orchards due to concerns about soil disturbance. Mowing, while ineffective at eliminating ryegrass, helps reduce seed production and slows weed proliferation.

To combat herbicide resistance, a collective management approach is necessary. Strategies focus on disrupting dispersal mechanisms to prevent the spread of resistant populations, reducing weed reproductive output, and implementing diverse, multitactic integrated weed management (IWM) programs. Reducing reliance on herbicides and promoting sustainable alternatives is essential, along with encouraging cost-effective nonchemical methods. Research underscores the need for sustainable weed management approaches tailored specifically for hazelnut orchards.

The primary research objective of this study is to evaluate nonchemical management tools for controlling and suppressing the reproductive capacity of Italian ryegrass (Fig. 9.3). Field studies were conducted in four locations in Oregon during the spring of 2023, including two sites in Corvallis, one in Shedd, and one in Mollala. These sites had high ryegrass infestation levels, with reports of herbicide escapees resistant to glyphosate, clethodim, and paraquat. The weed growth stage at the time of treatment ranged between 20–30 cm in the vegetative (boot) stage.

The experimental treatments included single- and double-pass mowing at a speed of 2 km h^{-1}, electrical weeding applied at 15 MJ/ha using the EH30 Thor Zasso device, and glufosinate at 1.68 kg a.i./ha, also applied as a single and double treatment. Combination treatments were also tested, where electric weed control (EWC) was followed by mowing or glufosinate, and vice versa. A nontreated control group was included for comparison.

Data collection and analysis involved evaluating weed control effectiveness at 28 and 56 days after the initial treatment (DAIT). Inflorescence density and biomass were measured at 56 DAIT, and seed viability and vigor were assessed at two of the study sites. The experiment was designed using a randomized complete block design (RCBD) with four replicates. Statistical analysis was conducted using a linear mixed model ANOVA with a significance level of $\alpha = 0.05$, and Tukey's honest significant difference test was applied for mean separation.

Results indicated significant differences between treatments and across experimental sites. Mowing alone was insufficient for complete suppression, with single-pass mowing having limited impact and double-pass mowing improving control but failing to eliminate ryegrass. Electrical weeding effectively reduced inflorescence density in three out of four sites, demonstrating comparable efficacy to herbicide electrical weeding. It also reduced inflorescence weight and overall reproductive output. Notably, electroweeding significantly decreased seed viability, indicating its potential as an alternative or complementary tool to chemical herbicides.

In conclusion, the study highlights the effectiveness of electrical weeding as a promising nonchemical weed management strategy, particularly in addressing herbicide resistance.

Effectiveness of Electrical Weed Control for the Organic Highbush Blueberry

The following is based on *Electrical Weed Control in Organic Highbush Blueberries* by Marcelo Moretti and Luisa Carolina Baccin, submitted to Oregon State University (Moretti and Baccin, 2024).

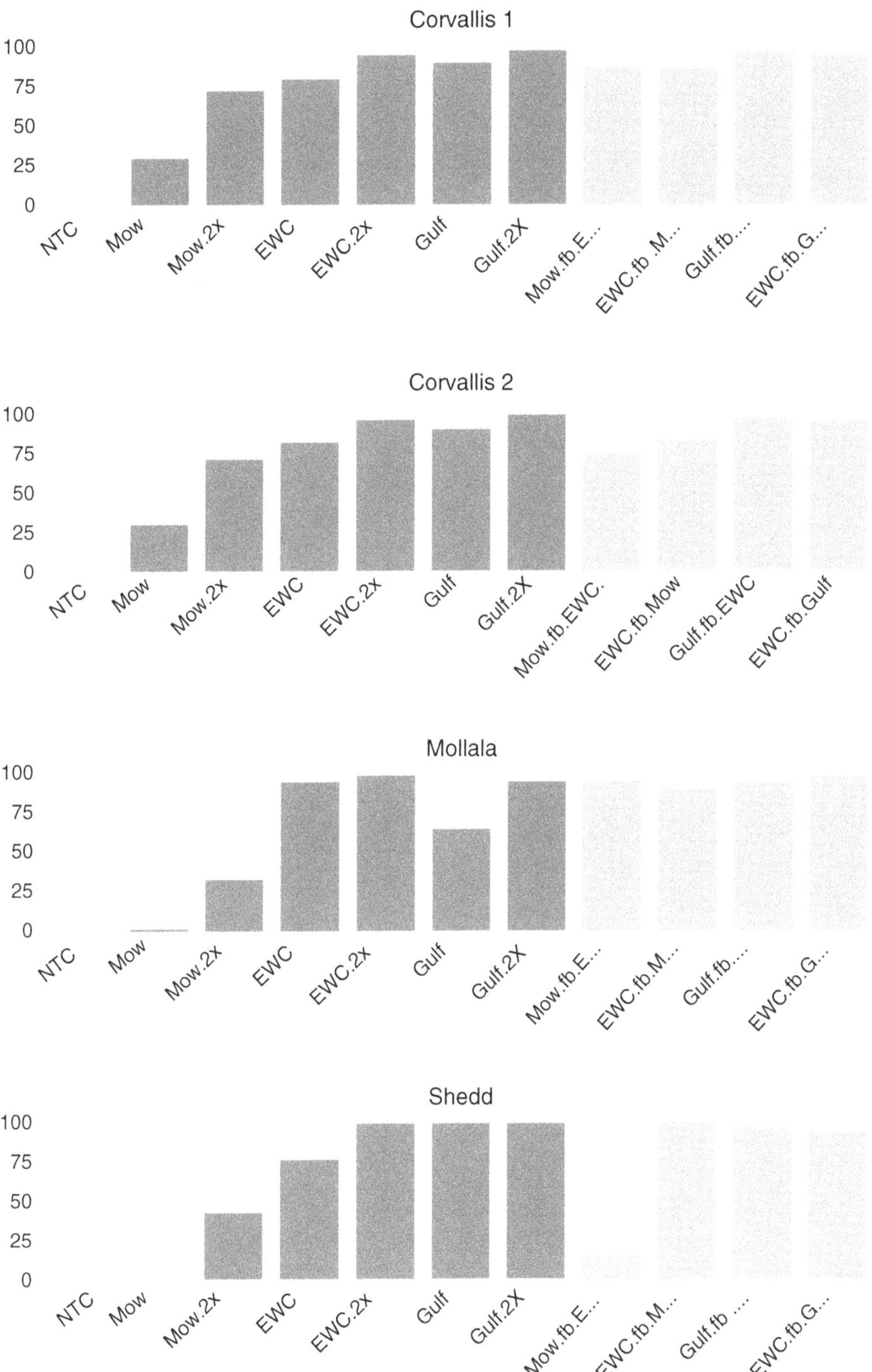

Fig. 9.3. Italian ryegrass control (%) results using different treatments (variation was low, therefore excluded from the original). (Reproduced with permission from Moretti (2022).)

Organic highbush blueberry production is expanding rapidly, particularly in the Pacific Northwest of the United States, where climatic conditions favor this high-value crop. However, one of the greatest challenges facing organic growers remains weed management, especially in systems where synthetic herbicides are prohibited. Perennial weeds such as Canada thistle (*Cirsium arvense*), yellow nutsedge (*Cyperus esculentus*), and field bindweed (*Convolvulus arvensis*) are especially problematic due to their deep and persistent underground structures.

EWC emerges as a promising nonchemical method to address these challenges. It operates by delivering high-voltage, high-current pulses to plant tissues, causing irreversible damage to cellular structures and leading to plant death. Its effectiveness is influenced by numerous factors, among which operational speed and the number of applications play critical roles.

Materials and methods

Field studies were conducted in certified organic blueberry farms across Oregon between 2022 and 2024. Two commercial EWC platforms—the Zasso XPS and Raiden systems—were deployed using tractor-mounted high-voltage applicators.

Five major experiments were designed:

- **Speed trials.** EWC was applied at varying speeds (0.5, 1, 2, 3, and 4 km h^{-1}).
- **Application frequency trials.** Comparisons between single vs sequential applications.
- **Species-specific response trials.** Focus on representative broadleaf and grass weeds.
- **Combined methods trials.** Integration with mowing.
- **Longitudinal trials.** Monitoring weed control over 28–56 DAIT.

Key data collected included visual weed control assessments, biomass measurements, and species-specific mortality at multiple time intervals post-application.

Results

Application speed directly influenced EWC efficacy. Slower speeds of 0.5–1 km h^{-1} delivered higher energy doses (up to 108 kJ m^{-2}), resulting in:

- 82–86% total weed control at 28 DAIT.
- Biomass reduction of up to 73% compared to untreated controls.

In contrast, as speed increased, total energy delivered per meter decreased. Each 1 km h^{-1} increment reduced efficacy by an average of 7–12%. At 4 km h^{-1}, weed control fell below 50% unless sequential applications were employed.

While single slow-speed applications were effective in the short term, long-term control improved with sequential treatments, even at higher speeds. Two applications at 2 or 4 km h^{-1} delivered up to 83% control at 42 DAIT, outperforming slower single applications whose effectiveness declined over time (e.g. 0.5 km h^{-1} dropped to 17% by 56 DAIT).

Weeds with softer tissue and thinner cuticles were more susceptible:

- *Epilobium ciliatum* and *Persicaria pensylvanica* saw >85–100% control at 35–42 DAIT.
- *Festuca arundinacea* and *Kickxia elatine* required higher energy (≥18 kJ m^{-2}) and multiple applications for moderate control (67–73%).

The effectiveness of EWC in organic blueberry production hinges on a nuanced balance between speed, energy delivery, and frequency of application. Slower speeds allow for deeper energy penetration, critical for damaging regenerative structures such as rhizomes and tubers. However, from a commercial perspective, extremely slow operations are impractical.

Sequential treatments at moderate speeds offer a more operationally efficient and agronomically effective strategy. This approach allows energy input per unit area to accumulate over time, weakening perennials and reducing their reproductive capacity without compromising field throughput.

Conclusion

EWC, when properly managed, offers a scalable and sustainable solution for organic blueberry systems. Key conclusions include:

- Lower operational speeds yield higher immediate weed control but have limited longevity.

- Sequential applications, even at moderate speeds, are essential for long-term efficacy.
- Species-specific responses necessitate adaptive management strategies.
- Integration with other mechanical practices (e.g. mowing) can enhance outcomes.

EWC represents a pivotal advancement in organic agriculture, enabling weed suppression without reliance on chemicals or excessive soil disturbance. With further optimization, it has the potential to become a cornerstone of IWM in perennial cropping systems.

Electrical Weed Management in Organic Highbush Blueberry Fields: Efficacy of Weed Control, Crop Response, and Soil Health

The following is based on *Electrical Weed Control in Organic Highbush Blueberries* by Marcelo Moretti and Luisa Carolina Baccin, submitted to Oregon State University (Moretti and Baccin, 2024).

Hereby is examined the efficacy of electrical weeding in organic highbush blueberry systems, focusing on three critical aspects: weed control effectiveness, crop response, and soil health.

Materials and methods

Experiments were conducted over two growing seasons (2023–2024) at an organic highbush blueberry farm in Corvallis, Oregon. Treatments included:

- EWC at two energy levels: 28 kJ m^{-2} (low) and 144 kJ m^{-2} (high).
- Mowing (control method).
- Mulch systems: bare ground, sawdust, and weed mat.

An RCBD was used with four replicates per treatment. Applications were timed to weed phenology and repeated biannually. Measurements were made related to:

- **Weed control.** Visual ratings (% control), dry biomass (g m^{-2}), species richness.
- **Crop response.** Canopy volume, shoot length, fruit bud set, stomatal conductance, and leaf nutrient content.
- **Soil health.** Physical (moisture, aggregation), chemical (pH, electrical conductivity [EC], NPK [nitrogen, phosphorus, potassium]), and biological (carbon dioxide [CO_2] respiration, microbial biomass, enzyme activity).

Results

Electrical weeding treatments significantly reduced weed biomass and species richness compared to mowing:

- **Biomass reduction.** High-energy EWC (144 kJ m^{-2}) reduced weed biomass by up to 72% across all mulch types.
- **Species shift.** Dominant annuals were eliminated, while perennial weeds (*Cirsium arvense, Convolvulus arvensis, Rumex crispus*) persisted but at lower densities.
- **Mulch interaction.** Weed mat combined with electrical weeding produced the highest control (~90%), followed by sawdust + EWC (~80%), and bare ground (~65%).

The consistent control over two seasons indicates that electrical weeding not only manages aboveground biomass but also suppresses regrowth through damage to belowground reproductive organs.

One of the most significant concerns with electrical weeding is its compatibility with crop performance, particularly given the sensitivity of blueberry roots to mechanical and electrical injury:

- **Canopy volume and shoot length.** No significant differences were found among weed control treatments, indicating that electrical weeding did not impede vegetative growth.
- **Fruit bud set.** Maintained across treatments; average bud set was 68% with no treatment interaction effects.
- **Leaf nutrient status.** Nutrient concentrations (N, P, K, Ca, Mg, Fe, Zn) remained within sufficiency ranges across all treatments, confirming no negative uptake effects.

Stomatal conductance, transpiration, and electron transport rate were measured to assess plant stress. Results showed no significant difference between mowing and EWC treatments, even at high energy doses, across mulch types. These results reinforce the suitability of EWC for organic systems, as it minimizes weed competition without detrimentally affecting plant physiology or nutrient uptake.

Sustainable weed control in organic farming must consider the long-term health of the soil ecosystem. To this end, the impact of electrical weeding on the soil's physical, chemical, and biological attributes was thoroughly examined:

- **Moisture and aggregation.** Electrical weeding treatments did not significantly alter gravimetric moisture or water-stable aggregates.
- **Soil pH and electrical conductivity (EC).** No significant changes were observed, indicating EWC did not influence soil ionic balance or salinity.
- **Macro- and micronutrients.** Total N, available P and K, and cation exchange capacity remained stable across treatments.
- **Soil respiration:** Short-term CO_2 evolution (24 h and 96 h) showed no significant differences, suggesting microbial metabolic activity was unaffected.
- **Microbial biomass and enzyme activity.** β-glucosidase activity and microbial biomass carbon levels remained consistent, confirming no suppression of beneficial microbial communities.

This aligns with previous studies indicating that, when properly applied, electrical weeding causes negligible disturbance to soil biota and function.

Conclusion

- EWC is a highly effective and nondestructive method for managing weeds in organic highbush blueberry systems.
- Weed suppression is most effective when EWC is used in combination with mulches, particularly weed mats and sawdust.
- EWC treatments did not negatively affect crop growth, physiology, or yield-related traits.
- Comprehensive soil health indicators showed no degradation, supporting long-term sustainability.

French Electrical Weeding Control in Potatoes

The following is based on "Défanage électrique: Nouvelles références" by Michel Martin, from Arvalis, Institut du Végétal (Martin, 2020).

In the constant search for sustainable agricultural practices, electric desiccation has emerged as a promising alternative to traditional chemical-based methods. The XPower Zasso system, a state-of-the-art electric defoliation technology, was rigorously tested in 2019 by Arvalis, Institut du Végétal at Villers Saint Christophe, France. This study aimed to assess its efficiency in different environmental conditions and on various potato varieties, considering factors such as soil moisture, mechanical grinding, and the use of chemical supplements. The field trials focused on two popular potato varieties, Challenger and Nicola, under a range of treatment scenarios (Fig. 9.4 and Fig. 9.5):

- **Mechanical grinding.** Some plots were preground before applying XPower, while others were left untouched.
- **Application speed.** The initial XPower pass was conducted at either 3 km h^{-1} or 5 km h^{-1}.
- **Reapplication timing.** In certain plots, a second pass was performed at 3 km h^{-1} exactly one week after the first.
- **Chemical supplementation.** Some plots received additional chemical treatments—either Beloukha (8l ha) or Spotlight (1l ha)—while others relied solely on XPower.
- **Watering before treatment.** A subset of plots was irrigated prior to the XPower application to evaluate the impact of soil moisture.

This detailed setup allowed for a thorough examination of XPower's performance across various conditions. To determine XPower's stand-alone efficiency, the study monitored regrowth patterns in different treatment conditions.

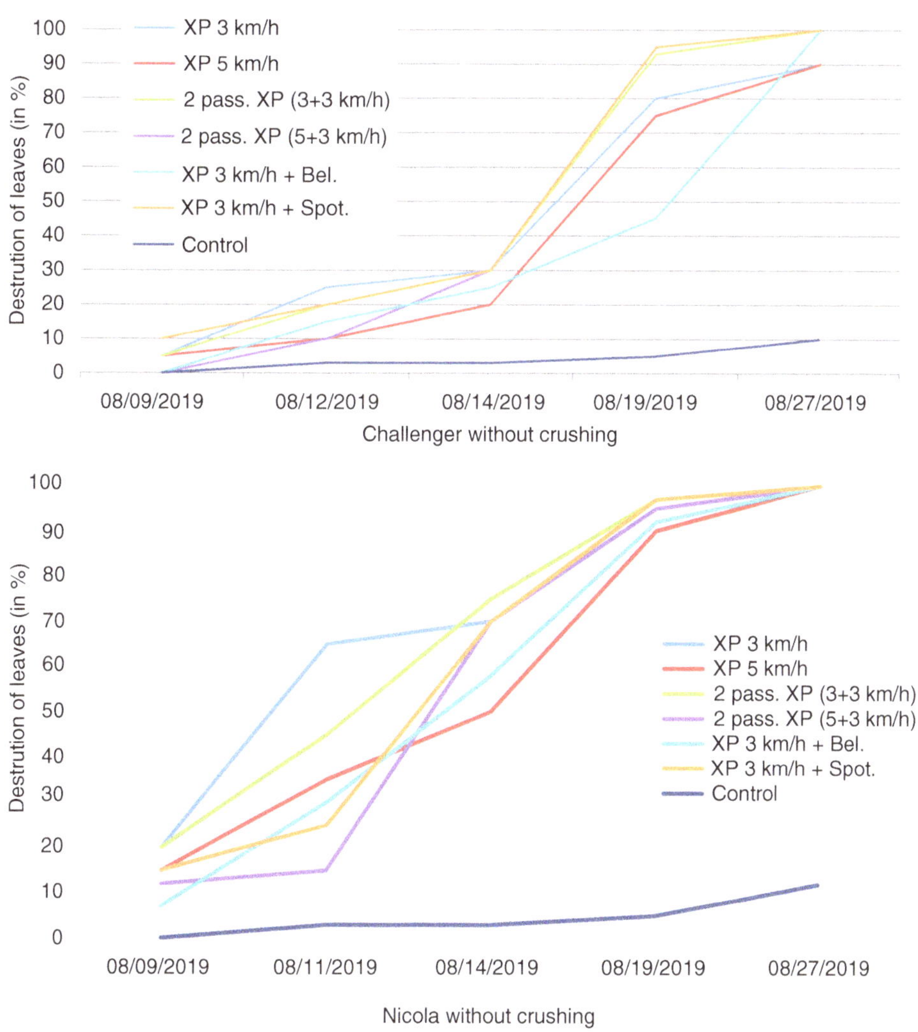

Fig. 9.4. Results of different electrical weeding treatments with previous mechanical treatment. (Reproduced with permission from Martin (2020).)

Regrowth in unground lots

A slight tendency toward more regrowth was observed in drier soil and unground plots.

Single-pass XPower treatments resulted in minimal regrowth, but certain harder-to-defoliate varieties exhibited greater persistence.

Regrowth after mechanical grinding

No regrowth was observed in preground plots treated with XPower, highlighting the benefits of combining mechanical and electrical desiccation.

One of the key objectives of the study was to evaluate whether a single application of XPower was sufficient to achieve effective defoliation.

- **Single pass.** While effective for easier-to-defoliate varieties, it struggled with more resilient plants.
- **Double pass.** A second application significantly improved defoliation efficiency, particularly for tougher potato varieties.

To ensure precise and reliable evaluations, the study incorporated drone imaging, which provided a comprehensive aerial view of the defoliation

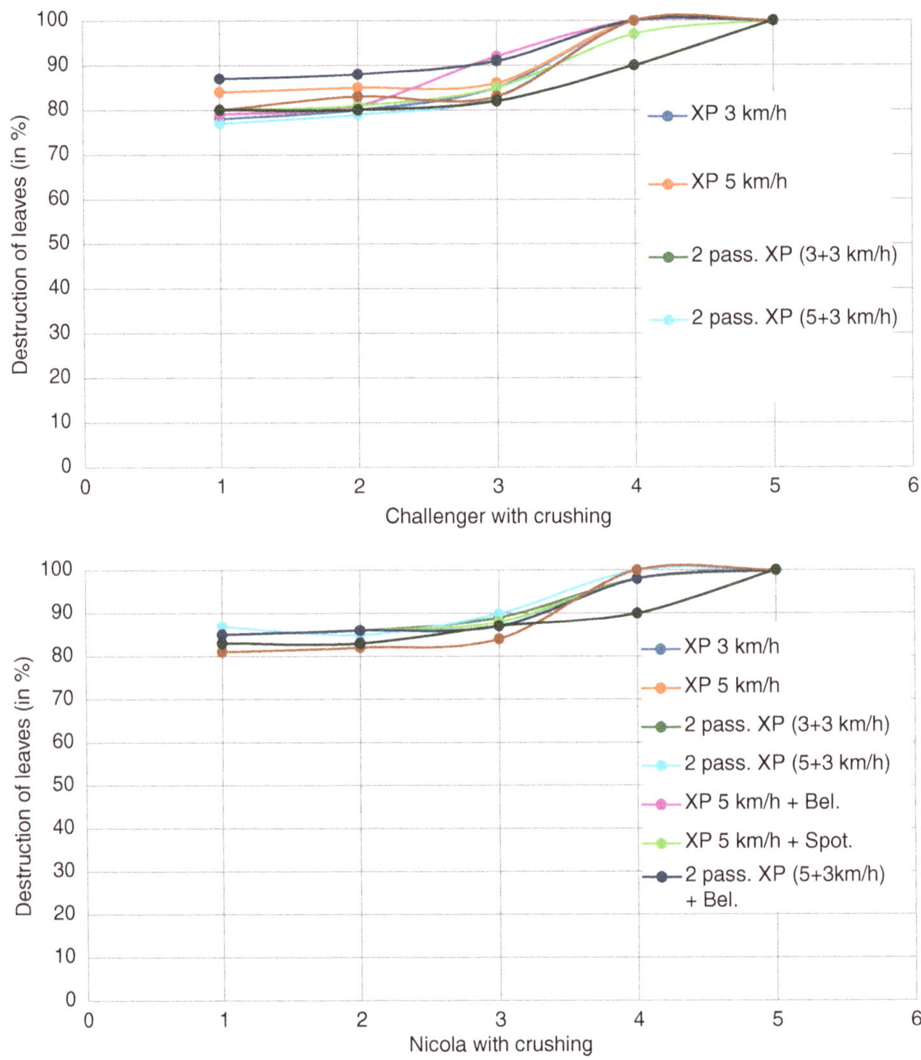

Fig. 9.5. Results of different electrical weeding treatments without previous mechanical treatment. (Reproduced with permission from Martin (2020).)

process. The key findings from the imaging analysis included:

- A strong correlation between drone observations and field evaluations.
- More distinct defoliation patterns in unground plots, reaffirming the influence of grinding.

The findings of the 2019 Villers Saint Christophe trial confirmed XPower Zasso as an efficient, chemical-free solution for potato defoliation. The primary conclusions were:

- XPower alone is effective for defoliation but requires a double pass for tougher varieties.
- Mechanical grinding enhances efficiency, allowing for a single XPower pass to achieve optimal results.
- Moist soil conditions slightly improve defoliation rates, though effectiveness remains high in dry conditions.
- Regrowth is minimized with grinding, eliminating the need for additional treatments.

- Drone imaging proved to be a valuable tool in assessing defoliation efficiency across large areas.

These results establish XPower Zasso as a viable alternative to chemical desiccants, offering an environmentally sustainable approach to potato cultivation.

Electrical Weed Control in Sugar Beet: A Comparison of Pre-emergence Methods

The following is based on *Electrical Weed Control in Sugar Beet: A Comparison of Pre-emergence Methods* by Maximilian Koch, Anastasia Hermann, Benjamin Ergas and Peter Risser from Zasso GmbH, Aachen, Germany, Universität Hohenheim, Stuttgart, Germany, and Südzucker AG, Geschäftsbereich Landwirtschaft, Versuchsgut Kirschgartshausen, Mannheim, Germany (Koch *et al.*, 2020).

Weed control in sugar beet cultivation is a costly and time-intensive process, with average expenses exceeding €300 per hectare, accounting for over 20% of total cultivation costs. Sugar beet plants are highly sensitive in their early growth stages, particularly before reaching the eight-leaf stage. To prevent yield losses, complete weed exclusion is necessary.

Historically, EWC in sugar beet cultivation has shown promise. Research conducted by Diprose *et al.* (1980) demonstrated the efficacy of electrical power in controlling weed beet and bolter populations. With increasing legal restrictions on synthetic chemical herbicides, growing herbicide resistance, and the push for more sustainable pesticide use, the demand for alternative weed control methods has never been greater. The application of Electroherb™ technology in sugar beet cultivation has been the subject of extensive field trials, including a pivotal study conducted at the experimental farm Kirschgartshausen by Südzucker AG in collaboration with Zasso GmbH in 2019.

The 2019 field trial at Kirschgartshausen aimed to assess the effectiveness of Electroherb™ as a nonchemical weed control method (Table 9.1). The experiment involved two pre-emergence treatments conducted 3 and 7 days after sowing. The treatments included mechanical weeding (harrow) and three different electrical weeding treatments (using 3 and 5 km h^{-1} speeds at 72 kW). Each treatment was applied to plots that had either received glyphosate pre-sowing or no chemical intervention.

Weed densities were evaluated 18 days after sowing across a 10 m^2 area. The results indicated that without glyphosate, the average weed density was 51 ± 35 plants per 10 m^2. With glyphosate application, weed density dropped to 14 ± 6 plants per 10 m^2. The mechanical treatment showed similar weed densities to untreated plots, both with and without glyphosate application. However, pre-emergence treatments utilizing Electroherb™ technology demonstrated significantly lower weed densities,

Table 9.1. Comparison of weed control strategies in sugar beet with focus on pre-emergence mechanical and electrical methods; here the "XPower," a prototype series of Electroherb™, was in action.

			Weeds per 10 m^2			
			With		Without	
Variant	Pre-emergence 1	Pre-emergence 2	Glyphosate		presowing	
1	–	–	14	a	72	a
2	Harrow 5 km h^{-1}	Harrow 4 km h^{-1}	21	a	72	a
3	–	–	17	a	54	a
4	–	–	13	a	88	a
5	–	–	23	a	97	a
6	XPower 3 km h^{-1}	–	16	a	5	b
7	–	XPower 3 km h^{-1}	6	a	7	b
8	XPower 5 km h^{-1}	–	5	b	11	b
Mean +- standard deviation			**14**	**±6**	**51**	**±35**
Date	3/25/2019	3/29/2019	4/19/2019			
Days after sowing	3	7	18			

particularly in plots without prior glyphosate application.

Variant 6, which combined Electroherb™ with glyphosate, resulted in an average of 16 weed plants per 10 m^2. This result suggested that new weed populations emerged between days 3 and 18 after sowing, particularly in the treatment area of this variant. No significant differences were observed between pre-emergence treatment one (3 days after sowing) and pre-emergence treatment two (7 days after sowing) regarding overall weed control.

Weed Control in Organic Soybean Using Electrical Discharge

The following is based on the "Weed control in organic soybean using electrical discharge" by Alexandre Magno Brighenti and Deodoro Magno Brighenti, from the Brazilian Federal University of Lavras and Embrapa (Brighenti and Brighenti, 2009).

With the rising demand for organic food since the 1980s, particularly in Brazil, organic farming has been expanding at an estimated annual growth rate of 20–30%. One of the major challenges faced by farmers transitioning to organic farming is weed management. The diversity of cropping systems necessitates ongoing research to develop innovative technologies that can effectively address weed control challenges while ensuring the sustainability of agricultural activities. This study evaluates the effectiveness of electrical discharge as a weed control method in organic soybean farming.

Experimental design

Two experiments were conducted in a no-till organic soybean field in São Miguel do Iguaçu, Paraná State, Brazil. An RCBD with four replications was used. The soybean variety "BRS 232" was sown on November 12, 2006, in 50 cm row spacing with plot dimensions of 4 m × 5 m.

- **Experiment 1.** Electrical discharge voltage set at 4,400 V.
- **Experiment 2.** Electrical discharge voltage set at 6,800 V.

Treatments

The treatments consisted of varying tractor engine speeds:

- 2,200 rpm (revolutions per minute);
- 2,000 rpm;
- 1,600 rpm;
- hand-weeded control;
- unweeded control.

The Electroherb™ device (Sayyou—Brasil Indústria e Comércio Ltda) was used to apply the electrical discharges. This equipment is attached to the tractor's power take-off, generating electricity and directing it through applicator cups mounted on a horizontal boom for precise weed targeting. The electrical discharge disrupts the physiological functions of weeds, causing them to wilt and die shortly after application. Weed control was applied only between the soybean rows.

Application and evaluation

The treatments were applied on December 15, 2006, when the soybean plants were at the V4 phenological stage. A Ford 6600 tractor was used at an average speed of 4 km h^{-1}.

The predominant weed species in the experimental area were:

- **Wild poinsettia** (*Euphorbia heterophylla*)
- **Morning glory** (*Ipomoea* spp.)
- **Prickly sida** (*Sida* spp.)
- **Alexandergrass** (*Brachiaria plantaginea*)
- **Crabgrass** (*Digitaria* spp.)

Weed density was approximately 88 plants m^{-2} in Experiment 1 and 36 plants m^{-2} in Experiment 2, with average weed heights ranging between 4–8 cm at the time of application. Control efficiency was evaluated on December 16, 2006 (1 day after treatment—1DAT) and on January 4, 2007 (20 days after treatment—20DAT). A percentage scale was used to assess weed control, where 0% represents no control and 100% indicates total weed mortality.

At preharvest, weed biomass was collected from a 0.25 m^2 sample area and dried at 65°C in a forced-air oven until a constant weight was reached. Soybean yield was recorded on March 23, 2007, and expressed in kilograms per hectare (kg/ha). Data were analyzed using ANOVA, and means were compared using Tukey's test (P≤0.05).

Efficiency

Experiment 1 (4,400 V)

At 1DAT, all three engine speeds resulted in statistically similar weed control levels. However, by 20DAT, 2,200 rpm provided 90% control, approximately 20% more effective than 1,600 rpm.

In terms of dry weed biomass, higher tractor speeds resulted in lower weed biomass, with 2,200 rpm showing the greatest suppression. The increase from 1,600 rpm to 2,200 rpm represented a 27% rise in engine speed, which directly influenced voltage output and improved weed control efficiency.

High-voltage application caused substantial cellular damage in weeds, with field applications of 170–330 W yielding effective control of grass weeds. Soybean yield was highest under the 2,200 rpm treatment, except for the hand-weeded control.

Experiment 2 (6,800 V)

As in Experiment 1, weed control at 1DAT was statistically similar across treatments, reaching approximately 90%. However, by 20DAT, 2,200 rpm achieved 100% control. The 2,000 rpm treatment also provided statistically equivalent control to 2,200 rpm and the hand-weeded control.

Visual observations showed that, in untreated plots, wild poinsettia (*Euphorbia heterophylla*) dominated, suppressing soybean growth. However, in the 6,800 V treatment with 2,200 rpm, weed control was effective, with only a few newly germinated plants appearing after application.

The higher voltage used in Experiment 2 resulted in improved weed control. Small weeds (4–6 cm height, 1–3 mm stem diameter) were effectively controlled with a single electrical discharge of 135 mJ, while taller weeds (80–120 cm, 10–15 mm stem diameter) required 15 kV for satisfactory control.

Diprose *et al.* (1980) found that controlling weedy beets in commercial beet crops required up to 20 kW to achieve complete mortality. Similarly, developed a portable device applying 3 kV, which effectively controlled *Poa annua* while ensuring operator safety has been developed.

Dry weed biomass was lowest under the 2,000 rpm and 2,200 rpm treatments in Experiment 2. Although no statistically significant differences were observed among the three tractor speeds, absolute soybean yield was highest at 2,200 rpm.

Author's comment

While the results of Experiment 2 (6,800 V) demonstrated improved weed control compared to Experiment 1 (4,400 V), it is important to highlight that this improvement was not solely due to the increase in voltage. Instead, the primary factor influencing weed control efficiency was the higher power applied during treatments.

Simply increasing voltage does not necessarily result in higher power unless there is a corresponding increase in power. In this study, the tractor engine speed directly influenced the electrical output of the system, affecting both voltage and current delivery. At a higher rpm (such as 2,200 rpm), the alternator was able to generate more electrical energy, leading to greater total power delivered to the weeds.

This distinction is crucial because it explains why higher rpm treatments consistently resulted in better weed suppression, regardless of voltage settings. While higher voltage can contribute to better penetration and electrical discharge efficiency, it is the total power applied that determines the extent of cellular damage in weeds, leading to their effective desiccation and mortality.

Therefore, the improved weed control observed in Experiment 2 is a direct consequence of higher power application translating into more energy per area, not just the increase in voltage from 4,400 V to 6,800 V.

Electrical Weed Control in Highbush Blueberries

The following is based on the 2024 publication *Electrical Weed Control in Highbush Blueberries* by Luiza Baccin and Marcelo Moretti, from the Department of Horticulture, Corvallis, Oregon, USA (Moretti and Baccin, 2024).

The pursuit of effective, sustainable, and organic weed control methods in highbush blueberry production has led researchers to explore innovative alternatives to conventional chemical herbicides. Among these, EWC emerges as a promising technique. By utilizing high-voltage electrical currents, this method disrupts plant cell functions, leading to tissue damage and root effects that hinder weed regrowth. This section delves into the efficacy, safety, and ongoing research concerning EWC in blueberry production.

Organic blueberry production and weed management challenges

The Pacific Northwest of the United States hosts a significant number of organic blueberry farms. Due to strict organic certification standards, traditional herbicides are not viable solutions, compelling farmers to seek alternative weed management strategies. Raised beds and weed mats are foundational techniques for cost-effective weed suppression, yet the growing interest in nonmechanical, nonchemical options has spurred research into EWC.

EWC operates by applying energy directly to weeds through electrodes. This energy is converted into heat within plant tissues, leading to irreversible cellular disruption. Several key factors influence the effectiveness of EWC, including plant species, plant size, soil moisture, soil conductivity, and overall plant density.

Research objectives and methodology

The current research on EWC in highbush blueberries focuses on two primary objectives:

- Evaluating crop safety and weed control efficiency:
  - Investigating the effects of EWC on different types of mulch, including bare ground, sawdust, and weed mats.
  - Assessing weed biomass reduction and ground coverage to determine the efficiency of EWC.
  - Studying potential impacts on blueberry plant health and yield.
- Control of invasive weed species:
  - Comparing EWC efficacy against conventional mowing techniques.
  - Evaluating EWC ability to suppress persistent weeds such as Canada thistle and yellow nutsedge.
  - Determining the effects of sequential applications on weed regrowth and seed viability.

Field trials conducted at the Lewis Brown Research Farm in Corvallis, Oregon, utilized varying energy levels and application speeds to assess performance under different conditions.

Findings and results

- Ground coverage and weed biomass reduction:
  - High-energy EWC applications significantly outperformed mowing in reducing weed biomass.
  - No notable differences in effectiveness were observed between different energy levels within the high-energy range.
- Impact on blueberry plants:
  - Blueberry plants showed no adverse effects from EWC treatments when applied correctly.
  - Mulch types influenced the outcomes, with sawdust providing additional insulation against unintended soil heating.
- Control of Canada thistle and yellow nutsedge:
  - Sequential applications demonstrated improved control over Canada thistle, reducing regrowth compared to single treatments.
  - Yellow nutsedge was more effectively managed with a combination of mowing followed by EWC applications (Fig. 9.6 and Fig. 9.7).

Conclusion and future directions

The integration of EWC into organic blueberry production presents a viable, sustainable alternative to mechanical and chemical weed management. Key advantages include reduced labor

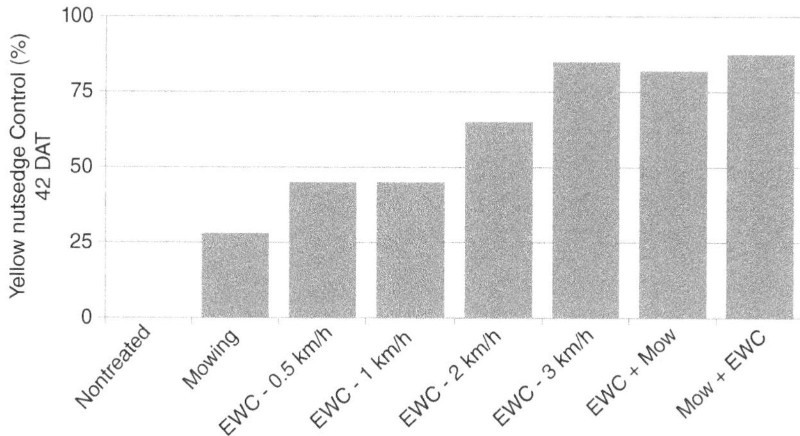

Fig. 9.6. Yellow nutsedge control (%) 42 days (variation was low and constant, therefore excluded from original). (Reproduced with permission from Moretti and Baccin (2024).)

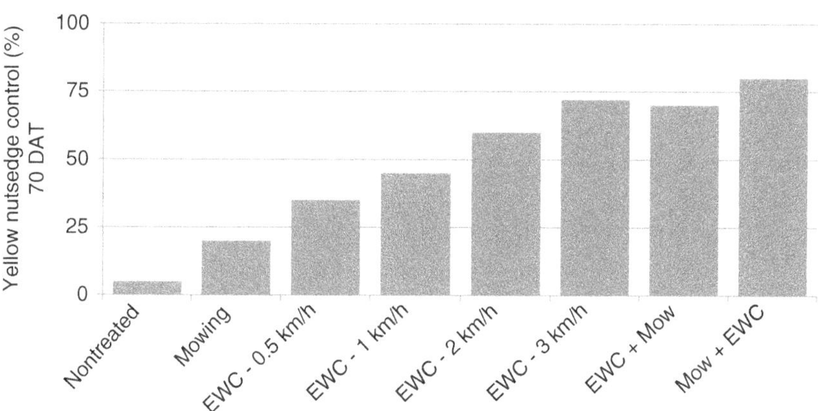

Fig. 9.7. Yellow nutsedge control (%) 70 days (variation was low and constant, therefore excluded from original). (Reproduced with permission from Moretti and Baccin (2024).)

inputs, long-term weed suppression, and compatibility with existing organic farming practices. Future research aims to optimize application parameters, explore cost-effectiveness, and refine best practices for large-scale implementation. As technological advancements continue, EWC holds the potential to revolutionize weed management in perennial fruit crops such as blueberries.

Cost Analysis and Profitability of Soybean Production Systems

The following is based on the 2012 publication *Cost Analysis and Profitability of Soybean Production Systems* by Danieli Simonetti, from IAPAR, Paraná, Brazil (Simonetti, 2012).

Soybean is an economically significant commodity in the international market, valued for its nutritional and functional protein content. It is cultivated extensively in the Southwest and West regions of Paraná, Brazil. This section presents a comparative analysis of four soybean production systems: transgenic, conventional, traditional organic, and organic with the use of the Electroherb™ technology. The study is based on data collected from various farming establishments, evaluating the financial, social, and environmental aspects of each system.

Objective and methodology

The primary goal of this study is to analyze and compare the profitability and economic viability of the different soybean production systems in the microregions of Capanema and Toledo for the agricultural year 2011–2012. Special emphasis is given to organic production systems, including traditional organic methods, and the use of Electroherb™, as well as their financial and sustainability implications.

The research employs a structured questionnaire for data collection during the 2011–2012 agricultural year. The sample comprises 21 farmers from Southwest region of Paraná and 3 from the West region. The distribution includes:

- Six farmers practicing traditional organic soybean farming.
- Six farmers employing the Electroherb™ system in organic soybean cultivation.
- Six farmers using conventional soybean farming methods.
- Six farmers utilizing transgenic soybean farming.

The data collected covers economic aspects of production, sustainability indices, and factors influencing the choice of production systems, as well as the challenges and advantages associated with each system.

- **Traditional organic farming (System A)**. Farmers in this group exhibit effective weed control practices, leading to reduced labor intensity and enhanced productivity. Some farmers diversify income sources by integrating additional produce, such as wheat and dairy production, for value-added revenue streams.
- **Organic farming with Electroherb™ (System B).** This system involves soybean cultivation with preplanting weed desiccation using the Electroherb™ technology. The method varies depending on weed infestation levels and individual farm conditions. Some farms experienced total crop failure and relied on agricultural insurance for compensation.
- **Conventional farming (System C).** Originating from the Green Revolution, conventional soybean farming relies on herbicides for preplanting desiccation and selective chemical applications for weed and pest control. This system reduces manual labor requirements and enhances efficiency.
- **Transgenic farming (System D).** The primary advantage of genetically modified soybeans is the ease of weed management, which minimizes manual labor and external workforce hiring.

The four systems can be broadly categorized into two groups: organic (A and B) and nonorganic (C and D). The Electroherb™ system is still in development, with expectations to reduce soil disturbance and improve weed control efficiency. Nonorganic systems show minimal variation in costs and productivity, with widespread adoption among farmers. However, there is significant consumer market resistance to transgenic soybeans.

The severe drought of 2011–2012 impacted all systems, causing production losses ranging between 30–100%. Despite cost fluctuations due to the drought, overall operational expenses remained relatively stable. The study found that organic production incurs lower costs due to the prohibition of herbicides and insecticides.

The key cost differentiators among the systems include expenditures on agricultural inputs and manual labor. In nonorganic farming, manual labor costs are nearly negligible, whereas organic farming requires more labor-intensive practices. Conversely, nonorganic agriculture incurs higher costs for chemical inputs.

Conclusion

Each soybean production system presents unique advantages and challenges. Traditional organic farming demonstrates sustainability benefits, albeit with higher labor demands. The Electroherb™ system shows promise in reducing soil disruption and improving weed management. Conventional and transgenic systems offer streamlined operations and lower labor costs but face resistance from market sectors favoring organic produce. Future advancements in weed management and sustainability strategies could further optimize soybean production across all systems. The largest aggregated value per hectare was obtained with the use of electrical weeding.

Effects of Soil Type, Contact Time, and Voltage on the Efficacy of Electrocution in Controlling Troublesome Weed Species

The following is based on the 2024 publication by Salvador and Pedroso (2024).

The widespread and intensive use of agrochemicals has led to herbicide-resistant weed biotypes, increasing control costs and complexity. In response, alternative strategies, such as electrocution, for weed control are being explored. This method shows promise in IWM, especially against resistant species.

Objective and methodology

To evaluate the effectiveness of electrocution as a method for weed control, considering its impact on different species and under various soil and operational conditions.

The study was divided into two distinct phases.

Phase 1: Controlled environment tests

- Weed species tested:
  - goosegrass (*Eleusine indica*);
  - sourgrass (*Digitaria insularis*);
  - Benghal dayflower (*Commelina benghalensis*);
  - castor bean (*Ricinus communis*).
- Soil types used:
  - medium-textured soil;
  - clay-textured soil.
- Variables tested:
  - electrode contact time;
  - shock intensity (two levels: tap3 and tap5, which refer to power settings on the electrocution equipment);
  - plants were exposed to electrical discharges, and their responses were visually monitored over time to determine the lethality of the treatment.

Phase 2: Field trials

- Variables tested:
  - generator power (shock intensity);
  - operational speed of the equipment during application.
- Evaluation methods:
  - visual control (observation of weed injury);
  - soil cover percentage, measured using Canopeo® (a mobile app for estimating green canopy cover)
- Timeline for evaluation:
  - Assessments were conducted at 3, 7, and 15 DAT to monitor the short- and medium-term:
  - effectiveness of the electrocution.

Conclusion

The results from both phases of the study show that electrocution is effective in controlling most of the tested weed species under a range of conditions. In particular:

- High control rates were achieved for goosegrass, sourgrass, and castor bean.
- Benghal dayflower exhibited significant tolerance to electrocution, suggesting species-specific resistance and the need for further study or adjustments in application parameters for such cases.
- In field trials, both lower speeds and higher electrical intensities (greater power from the generator) improved weed control outcomes.
- Despite these challenges, the study confirms that EWC is a viable alternative to chemical herbicides and offers a promising tool for IWM, particularly in urban environments, where localized and precise applications are advantageous.

References

Brighenti, A.M. and Brighenti, D.M. (2009) Controle de plantas daninhas em cultivos orgânicos de soja por meio de descarga elétrica. *Ciência Rural* 39(8): 2315–2319.

Diprose, M.F., Benson, F.A. and Hackam, R. (1980) Electrothermal control of weed beet and bolting sugar beet. *Weed Research* 20(5): 311–322.

Hamberg, R. (2025) Insights on inter-row electrical weeding as a non-chemical weed management tool in organic cotton. Texas A&M AgriLife Organic, College Station, TX, USA. Available at: https://agrilifeorganic.org/2025/07/23/insights-on-inter-row-electrical-weeding-as-a-non-chemical-weed-management-tool-in-organic-cotton/ (accessed November 10, 2025).

Koch, M., Hermann, A., Ergas, B. and Risser, P. (2020) *Electrical Weed Control in Sugar Beet: A Comparison of Pre-Emergence Methods*. Universität Hohenheim, Stuttgart, Germany.

Martin, M. (2020) *Le défanage électrique, une technique qui fait ses preuves*. ARVALIS – Institut du végétal, Beaucouzé, France.

Moretti, M.L. (2022) *Unique Way to Take Charge of Weed Control in Hazelnut Crops*. Oregon State University College of Agricultural Sciences, Corvallis, OR, USA.

Moretti, M.L. and Baccin, L. (2024) *Electrical Weed Control in Highbush Blueberries*. Oregon State University Department of Horticulture, Corvallis, OR, USA.

Salvador, M.A. and Pedroso, R.M. (2024) *Effects of Soil Type, Contact Time, and Voltage on the Efficacy of Electrocution in Controlling Troublesome Weed Species*. Escola Superior de Agricultura "Luiz de Queiroz" (ESALQ/USP).

Simonetti, D. (2012) *Cost Analysis and Profitability of Soybean Production Systems*. IAPAR, Paraná, Brazil.

10

Development Outlook: The New Frontiers in Electric Weeding

Weed Density Mapping System Using Drones

This is a general description of a weed density mapping system using drones. More specific solutions and practical applications will be explored throughout the chapter.

Weed management is a critical aspect of modern agriculture. Electrical weeding, a nonchemical and sustainable alternative, requires precise information about weed location and density to optimize its effectiveness. This section outlines a comprehensive approach to weed density mapping using drones (unmanned aerial vehicles [UAVs]), and integrating data acquisition, processing, analysis, and application in electrical weeding operations.

Examples of components

RGB cameras are used to capture high-resolution imagery, providing clear and detailed visual information about the field. These images help in identifying surface features and are essential for creating accurate orthomosaics and baseline maps of the area.

Multispectral and hyperspectral sensors go beyond the visible spectrum, capturing data across multiple wavelengths. This allows for the calculation of vegetation indices such as the normalized difference vegetation index (NDVI) and the green normalized difference vegetation index (GNDVI), which are critical for distinguishing between healthy crops and weeds based on their spectral signatures.

Thermal cameras detect heat differences across the field surface, offering an additional layer of contrast that can be particularly useful in certain scenarios where weeds exhibit distinct thermal characteristics compared to the surrounding vegetation or soil.

Control system

Effective weed mapping with drones relies not only on aerial hardware and imaging sensors but also on a robust and responsive ground control system. This system serves as the operational backbone, ensuring that data acquisition is precise, consistent, and aligned with agronomic goals. From initial mission planning to real-time adjustments during flight, the ground control infrastructure allows for seamless coordination between human operators and autonomous aerial platforms. Its design must account for the challenges of varied terrain, weather variability, and the need for precise georeferencing.

DOI: 10.1079/9781836992288.0010

The ground control system plays a crucial role in managing and coordinating the drone's operations during weed mapping missions. It includes real-time flight planning and control software that allows operators to define flight parameters such as altitude, speed, and the path over the area of interest. This software is essential for ensuring complete coverage of the target field and for achieving the necessary image overlap for later processing.

In addition to planning and control, the ground system also manages the communication link between the drone and the operator. This link transmits telemetry data in real time, including information on the drone's location, battery status, sensor performance, and flight conditions. It also supports remote monitoring, enabling operators to track the mission's progress and make immediate adjustments if needed. Together, these components ensure the safe, efficient, and precise execution of drone-based weed mapping operations.

In summary, the ground control system is indispensable to the success of UAV-based weed mapping. It provides the interface through which flights are planned, executed, and monitored in real time, ensuring both data quality and operational safety. By maintaining continuous communication with the drone and offering flexible in-flight management, the system lays the foundation for reliable and repeatable weed detection missions across diverse agricultural environments.

Data storage and onboard processing

As drones carry out weed mapping missions, they generate large volumes of high-resolution image and sensor data that must be stored, managed, and in some cases processed in real time. The efficiency and effectiveness of this data handling depend significantly on the drone's onboard storage capacity and computational capabilities, especially in field conditions where connectivity and bandwidth may be limited.

Typically, drones are equipped with onboard storage solutions such as SD cards or solid-state drives (SSDs). These components are essential for saving raw imagery and telemetry data collected during the flight, preserving it for later transfer to ground-based systems where advanced analysis and map generation can be performed. The choice between SD cards and SSDs often depends on the size of the datasets expected and the required write speeds, with SSDs generally offering higher capacity and faster data handling.

In addition to storage, some drone systems incorporate edge computing devices, such as NVIDIA Jetson modules. These compact, high-performance processors enable the drone to perform preliminary data processing directly on board during flight. This might include tasks like compressing image data, computing vegetation indices, or even running machine-learning models for real-time weed detection. Edge computing reduces the need to transmit large files in real time and can accelerate decision-making in adaptive weeding systems.

The integration of efficient storage systems and optional onboard computing represents a significant advancement in drone-based weed mapping. These technologies allow drones to operate with greater autonomy and adaptability, ensuring that high-quality data is not only captured but also intelligently managed, even before it reaches post-processing workflows on the ground. As data demands grow, such onboard capabilities will become increasingly vital to support scalable and responsive precision agriculture solutions.

Flight planning and data collection

Accurate weed density mapping using drones begins with careful preparation and adherence to rigorous data collection standards. The planning phase determines the quality and usefulness of the imagery captured, which in turn affects the precision of the weed detection algorithms and the resulting prescription maps. Both flight planning and in-flight data acquisition must be strategically designed to ensure consistency, coverage, and geospatial reliability.

Flight planning starts with defining the area of interest using georeferenced boundaries. This ensures that the drone's flight path aligns precisely with the field regions requiring weed monitoring. The next step is to determine the appropriate flight altitude and trajectory to achieve the desired ground sampling distance, which directly influences the resolution and clarity of the images. Careful consideration is given to flight path overlap, typically set at 80% forward and 60% side overlap. This level of redundancy is necessary to support image stitching and three-dimensional surface modeling, ensuring that all areas are thoroughly covered and minimizing gaps in the dataset.

Once the flight path is programmed, data collection follows a strict protocol. Flights are conducted during stable lighting and favorable weather conditions to reduce shadows, glare, and image distortion. This ensures that vegetation signatures remain consistent throughout the dataset. During each flight, the drone captures synchronized multispectral and RGB imagery, which is essential for generating layered datasets used in vegetation index analysis and object classification. To further enhance spatial accuracy, ground control points are strategically placed and surveyed throughout the field. These reference markers allow for precise correction of the drone's imagery during processing, improving the reliability of spatial measurements and map outputs.

In conclusion, flight planning and data collection form the foundation of a reliable drone-based weed mapping system. By carefully defining the flight parameters and adhering to a structured data acquisition protocol, operators can ensure that the resulting imagery is both high in resolution and rich in spatial accuracy. These early steps are critical to achieving accurate weed detection, generating effective density maps, and, ultimately, supporting more efficient and targeted electrical weeding strategies.

Introduction to image processing and weed detection

After raw imagery is collected from drone flights, it must undergo a series of processing steps to transform it into meaningful, actionable data. This transformation enables the identification and classification of weeds within a field, supporting precise and efficient electrical weeding interventions. The image-processing workflow consists of three main stages: preprocessing, vegetation index computation, and weed detection and classification.

The preprocessing stage begins with stitching together the individual images captured during the flight. These images are merged to create a seamless, georeferenced orthomosaic that covers the entire area of interest. To ensure data consistency and accuracy, radiometric and geometric corrections are applied. Radiometric correction adjusts for variations in light conditions and sensor response, while geometric correction aligns the imagery to real-world coordinates, compensating for lens distortion and perspective effects. Additionally, reflectance calibration is carried out using ground-based calibration panels. These panels, placed within the field before the flight, provide known reference values that allow for standardization of image brightness and contrast, ensuring that vegetation indices derived from the images are accurate and comparable across time and locations.

Once the imagery is preprocessed and calibrated, vegetation indices are computed to differentiate crops from weeds. Common indices such as the NDVI, GNDVI, normalized difference red edge (NDRE), and soil-adjusted vegetation index (SAVI) highlight differences in plant health, biomass, and species based on spectral reflectance properties. Histogram analysis is then used to identify distribution patterns of vegetation index values, which helps in setting dynamic thresholds to distinguish between target crops and unwanted weeds in varying field conditions.

With vegetation highlighted, the next step is to detect and classify weeds within the imagery. This process can follow two main approaches. In supervised classification, machine-learning models, such as support vector machines or random forest classifiers, are trained on manually labeled samples of crops and weeds. These models learn to identify and categorize different vegetation types based on spectral and spatial characteristics. Alternatively, deep learning approaches use convolutional neural networks (CNNs), such as U-Net or Mask R-CNN, to perform pixel-wise segmentation of weeds. These models are capable of capturing complex patterns and shapes, allowing for more precise detection, especially in dense or overlapping vegetation.

Following classification, post-processing techniques are applied to refine the outputs. Morphological operations, such as erosion and dilation, help clean up the segmented images by removing noise and connecting fragmented weed regions. Finally, spatial clustering algorithms group weed detections into defined density zones, which can be used to create weed maps that guide variable-rate electrical weeding applications.

The image-processing pipeline transforms raw drone imagery into precise weed classification maps through a combination of technical corrections, vegetation analysis, and advanced modeling techniques. By systematically applying these methods, it becomes possible to identify and quantify weed populations with high accuracy. This information is essential for creating targeted treatment strategies, ensuring that electrical weeding equipment applies power where it is most needed while conserving energy and protecting crops.

Weed density mapping and integration with electrical weeding equipment

Once weeds have been accurately identified and classified from drone imagery, the next critical step is the generation of weed density maps. These maps serve as the foundation for site-specific weed management strategies, enabling electrical weeding equipment to target infestations with precision and efficiency.

The generation of weed maps begins by compiling the results of weed classification into a raster format. This raster can represent either the probability of weed presence at each pixel or a simpler binary classification that indicates the confirmed presence or absence of weeds. This data is then processed and converted into vector shapefiles, which are widely used in agricultural machinery and geographic information systems. These vector files allow for seamless integration with field equipment and serve as the spatial reference for subsequent operations.

After the weed areas have been identified, the next stage involves density zonation. This step categorizes the weed-affected regions into distinct tiers, typically labeled as low, medium, or high density. Such stratification helps tailor the intensity of treatment to the severity of the infestation. The resulting data is exported in formats such as geoTIFFs and shapefiles, which are compatible with precision agriculture platforms and enable automated responses from electrical weeding machines in the field.

With the density zones defined, the system then moves into integration with electrical weeding equipment. The first step in this process is the creation of prescription maps. These maps are derived from the weed density data and are formatted into machine-readable instructions that define which areas of the field require treatment and at what intensity. To ensure safety and application accuracy, spatial buffers are applied around treatment zones, and margins are set to accommodate equipment width and field variability.

Once the prescription map is uploaded, the weeding equipment dynamically adjusts its electrical output based on the mapped weed density. In denser areas, the system may increase voltage, extend pulse duration, or modify electrode configurations to ensure thorough treatment. Conversely, in sparsely infested zones, it reduces the intensity to conserve energy and minimize unnecessary soil impact. These settings are calibrated in line with the specifications and operational parameters of Zasso equipment, ensuring that voltage levels, treatment time, and electrode behavior are all optimized for each application scenario.

Finally, real-time synchronization with a global navigation satellite system (GNSS) ensures that the electrical weeding system applies treatment precisely in the mapped zones. The equipment continuously references its position against the prescription map, activating the weeding system only where weeds have been detected. As it operates, the system logs its movements and treatment activities, creating a digital record of weed control events that can be used for further analysis and compliance reporting.

The process of weed density mapping and its integration with electrical weeding machinery creates a closed-loop precision agriculture system. By linking detailed spatial data with real-time, variable-rate treatment, this approach maximizes weed control effectiveness while minimizing energy use and environmental disturbance. It also lays the groundwork for adaptive and intelligent weeding systems capable of responding dynamically to changing field conditions and weed pressures.

Integration with electrical weeding equipment and system validation

The effectiveness of drone-based weed detection is ultimately realized when the generated data is integrated into electrical weeding systems. This integration allows for precise, targeted weed control that optimizes energy usage and improves overall treatment outcomes. The process begins with the transformation of weed density maps into actionable formats and continues through real-time execution in the field.

The first step in this integration involves the creation of prescription maps. These maps convert the visualized weed density data into a format that electrical weeding machines can interpret and act upon. The prescription maps divide the field into zones based on the density and distribution of weeds, specifying where and how the equipment should apply treatment. To ensure safety and precision, spatial buffers and margins are incorporated into the prescription. These adjustments account for factors such as equipment size, turning radius, and field irregularities, ensuring that electrical discharge is applied only within the intended treatment areas.

Following prescription creation, the electrical weeding system adapts its output based on the zoned data. This variable power application is designed to match the intensity of the treatment to the severity of the infestation. Areas with high weed density receive stronger or longer electrical pulses, while zones with sparse or no weeds receive minimal or no treatment. These adjustments are directly linked to the operational parameters of Zasso equipment, which include setting the appropriate voltage levels and pulse durations, and selecting the correct type of electrodes. The goal is to deliver just enough power to effectively eliminate weeds while minimizing energy waste and potential soil disruption.

The application of this treatment in the field is synchronized with real-time GNSS data. As the equipment moves across the field, its location is continuously compared against the prescription map. Treatment is activated only when the machine enters a predefined weed zone, ensuring that power is not applied in areas without infestation. Throughout the operation, the system logs positional data and treatment activity, producing a digital record of where and how the field was treated. This information is vital for monitoring performance, ensuring regulatory compliance, and planning future interventions.

To ensure that this integrated system functions accurately and effectively, rigorous calibration and validation protocols are required. Field validation begins with ground truth sampling, often using quadrant-based sampling methods to collect real-world weed presence data at specific locations. These samples are then compared with the weed detection results from the drone imagery. The performance of the mapping algorithm is evaluated using metrics such as the confusion matrix, and precision, recall, and overall accuracy. These statistical tools help determine how well the system identifies true weed locations versus false positives or negatives.

In addition to mapping accuracy, it is essential to evaluate the efficacy of the electrical weeding itself. This involves revisiting treated zones after application and assessing weed mortality. The results are correlated with the initial weed density classifications to determine whether the correct amount of electrical power was applied in each zone. If discrepancies are observed, such as underperformance in areas that were mapped as high density, the mapping algorithm and classification thresholds are refined accordingly. This feedback loop ensures continuous improvement in the system's performance and reliability.

The seamless integration of weed density maps into electrical weeding systems represents a significant step forward in precision agriculture. By translating spatial data into targeted treatment actions, these systems reduce unnecessary energy expenditure and enhance weed control efficacy. The calibration and validation processes ensure that both the detection and treatment phases are grounded in field reality, enabling adaptive improvements over time. Together, these components form a closed, intelligent loop that delivers sustainable and highly effective weed management.

System calibration and validation

A crucial phase in deploying a drone-based weed mapping system for electrical weeding is the calibration and validation of its performance.

This stage ensures that the data produced from aerial imagery accurately reflects the conditions on the ground and that the electrical treatment applied based on this data is effective. Without proper validation, even the most advanced mapping algorithms risk misclassification or inefficient resource application.

Field validation begins with ground truth sampling. In this method, a series of predefined quadrants are established throughout the treatment area, usually selected to represent a mix of weed densities and crop conditions. Within each quadrant, weed presence is manually recorded and classified. These observations are then compared to the corresponding areas on the drone-generated weed maps. This comparison allows for an assessment of the mapping system's ability to detect and classify weed populations accurately.

To quantify the performance of the mapping algorithms, statistical tools such as the confusion matrix are used. This matrix helps to identify the number of correct and incorrect classifications by the model, distinguishing between true positives, false positives, false negatives, and true negatives. From this, precision and recall metrics are calculated. Precision indicates how many of the weeds identified by the system were actually weeds, while recall shows how many of the real weeds present in the field were successfully detected. Together, these metrics offer a clear picture of the system's reliability and highlight areas for potential refinement.

Beyond detection accuracy, the system must also be validated in terms of treatment efficacy. This involves correlating the mapped weed zones with actual weed mortality after electrical weeding has been applied. Operators return to the same field locations posttreatment to evaluate whether weeds identified as needing treatment were effectively neutralized. If areas with high weed density show low mortality rates, it may indicate that the electrical power applied was insufficient or misaligned with the weed characteristics in those zones. Conversely, areas where weeds were eliminated without prior detection may suggest false negatives in the mapping process.

The final step in the validation loop is adjustment. If discrepancies are observed between the predicted weed presence and treatment outcomes, adjustments are made to the weed mapping algorithms. This could involve refining classification thresholds, retraining machine-learning models with additional labeled data, or integrating new vegetation indices to improve detection under specific crop or weed conditions. This iterative feedback process ensures continuous improvement in both mapping accuracy and treatment effectiveness.

System calibration and validation are essential for the success of drone-integrated electrical weeding operations. By grounding the mapping outputs in real-world observations and measuring the effectiveness of treatment outcomes, the system evolves into a more precise and dependable tool. This validation process not only reinforces confidence in the technology but also drives ongoing refinement, ensuring that each season and field condition is met with improved accuracy and greater weed control efficiency.

Conclusion

The integration of drone-based weed mapping with precision electrical weeding marks a transformative advancement in sustainable agriculture. By leveraging high-resolution imagery, advanced data processing, and machine-learning techniques, this system enables the accurate detection, classification, and mapping of weed populations across diverse field conditions. These insights are then seamlessly translated into actionable prescription maps that guide variable-rate electrical treatment, allowing for targeted weed control that minimizes energy use and avoids damage to crops or soil.

Equally important is the system's ability to self-improve through rigorous calibration and validation protocols. Ground truth sampling and statistical analysis ensure the reliability of weed detection, while posttreatment efficacy assessments feed back into algorithm refinement. This closed-loop framework fosters continuous learning and adaptation, making the system increasingly effective with each application cycle.

As regulatory pressures mount against chemical herbicides and the need for environmentally responsible practices intensifies, such precision-driven, nonchemical solutions offer a clear path forward. The combination of UAVs, artificial intelligence (AI), and electrical weeding technologies is not just a technical innovation—it is a practical, scalable, and ecologically sound approach to modern weed management. With ongoing advancements in sensor technologies, real-time processing, and autonomous systems, the potential for further optimization is vast, paving the way for smarter, cleaner, and more efficient agricultural practices.

Green-on-Brown Sensing

In modern agriculture, the ability to detect and eliminate invasive weeds is increasingly important. Invasive weeds can severely harm crop yields by competing for nutrients, water, and sunlight, and they often harbor pests and diseases. Advanced sensing technologies—including satellite imagery, drones, ground-based sensors, and sophisticated image-processing algorithms—have emerged to identify weed infestations quickly and accurately. These technologies enable farmers to target weed control measures only where needed, minimizing unnecessary herbicide use and reducing environmental impact (Zasso and SENAI, 2024).

Despite these advancements, many existing weed detection systems face practical limitations. Some solutions rely on remote cloud processing and wireless communication, which can introduce signal transmission failures and latency. For example, one patented "smart weeding" device sends captured images to a cloud server for processing and then uses radio-frequency signals to trigger weeding actions—an approach prone to communication failures between the camera sensor and the weeding equipment. Other approaches use complex image analysis algorithms that demand high computational power. One system uses a "Super Green" index (essentially, an excess-green formula 2G-R-B to highlight green vegetation) combined with Otsu's thresholding method for segmentation. While effective at isolating green pixels, that design still requires a powerful processor and omits any noise-reducing convolution step, leading to suboptimal results. Another prior solution employs an autonomous machine vision system to classify plant types and decide if a plant is a weed to eliminate, but it demands very high processing power and storage, making the equipment expensive and bulky. Yet another system focuses on counting intentionally planted crops using a camera between crop rows, rather than directly targeting weeds; it too requires high computing capacity and is cost-prohibitive for widespread use. In summary, existing weed sensing methods often suffer from one or more of the following drawbacks:

- **High complexity and cost.** Many systems use computationally intensive algorithms or cloud-based processing, necessitating expensive hardware or reliable network connectivity. This increases the cost and complexity of deployment in the field.
- **Communication reliability issues.** Wireless links (e.g. radio-frequency triggers) between sensors and weeding actuators can fail or suffer interference, potentially missing weed targets.
- **Insufficient noise filtering.** Some algorithms do not adequately filter out noise (e.g. small false-positive green pixels), as they skip steps like convolutional smoothing. This can lead to either false triggers or the need for higher processing power to compensate.

The objective of the proposed innovation is to address these limitations, and a new integrated sensing system and method for identifying invasive plants has been developed. The system is designed to reliably detect weeds by their green coloration using efficient onboard image processing, and to directly trigger weed elimination equipment in real time. This section describes the system's technical design—including its architecture, image-processing workflow, and operation—and explains how it improves upon prior art in efficiency, cost, and reliability.

Technical background and system overview

The invasive plant sensing system operates as a self-contained module that can be mounted on agricultural machinery (such as a weeding machine). The core components include a charge-coupled device (CCD) camera (color RGB) for image acquisition and a microcontroller for onboard image processing. The camera is interfaced to the microcontroller via a high-speed parallel camera interface (digital camera memory interface; DCMI) which allows direct transfer of image frames into memory. The microcontroller processes each image frame to detect green-colored regions corresponding to weeds, and if a weed is detected, it issues a binary output signal to activate a connected tool on the machinery.

Unlike systems that transmit images to an external computer or cloud, this invention performs all processing locally (on an embedded microcontroller). The device achieves a frame capture and processing rate of about 10–15 frames per second (FPS), enabling near-real-time response as the machine moves through the field. As soon as a significant green region is identified in a frame, the microcontroller sends a trigger signal via an RS-485 interface to the machine's control module. RS-485 is a robust wired serial communication standard often used in agricultural and industrial equipment for its noise immunity and reliability. Using a wired link (instead of wireless) to send the trigger virtually eliminates communication failures and interference when activating the weeding mechanism.

The output from the sensor system is a binary signal (on/off) that indicates the presence of a weed. In a typical configuration, the signal is a pulse of a fixed duration (e.g. ~300 ms). This pulse is fed into the machine's power module, which drives the weeding tool. For example, if the machine is an electric weeder, the power module would discharge an electric current into the soil at the location of the detected weed during that pulse. In other implementations, the triggered tool could be a sprayer, mechanical cutter, or any weed control actuator—the sensor system is agnostic to the type of actuation, as it simply provides a trigger signal when a target is detected. Power for the sensing system is supplied by the host machine.

The machine's battery (or power supply) feeds a power conditioning circuit that protects against surges and then provides regulated voltages to the camera and microcontroller. This ensures the sensing module operates within proper voltage levels, even in harsh conditions. The microcontroller also interfaces with auxiliary systems on the machine. For instance, it can control light-emitting diode (LED) illumination on the machinery based on ambient light conditions—using the average luminance of the captured image (Y_avg) to adjust LEDs—so that the target area is well-lit for the camera. This adaptive lighting helps maintain detection accuracy from dawn to dusk.

In summary, the system's architecture centers on a camera–microcontroller pair tightly integrated with the machine's hardware:

- **Camera (vision sensor).** A CCD RGB camera captures images of the ground ahead. It connects to the microcontroller via a DCMI interface for fast data transfer.
- **Microcontroller (processing unit).** The microcontroller receives image frames and runs the weed detection algorithm in real time. It also communicates with the camera and with the machine's actuators.
- **Power module and actuator.** The power module on the machine receives the RS-485 trigger and drives the weeding tool—e.g. firing an electric weeder or activating a sprayer—for a controlled duration.
- **Power supply and protection.** A power source, typically the machine's battery, supplies the system. Protective and regulatory circuits condition this supply for the camera and processor.
- **Communication link examples.** DCMI (camera to microcontroller data), I²C (camera control), and RS-485 (microcontroller to machine) are used for high reliability and simplicity, avoiding wireless transmission pitfalls.

With the hardware in place, the novelty of this invention lies in the image-processing method and how it efficiently detects green weeds with minimal processing overhead. The following section details the step-by-step method that the microcontroller executes for each frame.

Image-processing workflow

The weed detection algorithm processes each captured frame through a sequence of stages. Broadly, these stages are as follows: image capture, preprocessing, processing, and result. These stages transform the raw camera data into a binary decision (weed present or not) which then drives the output signal. Below is a breakdown of each step in the pipeline.

Image capture

The camera captures an image of the field scene. In this system, the camera outputs frames in a compact format to reduce data size.

Frame transfer and preprocessing

The image frame is transmitted to the microcontroller over the DCMI interface. This preprocessing stage ensures that the image is in a standard format and maximizes color fidelity for analysis.

Color space conversion (RGB to hue, saturation, value)

To detect green vegetation reliably, the system transforms the image from the RGB color space into the hue, saturation, value (HSV) color space. The HSV representation separates the color information (hue) from color purity (saturation) and color intensity (value), which is very useful for color-based segmentation. The conversion is done pixel by pixel, using standard formulas derived from the RGB color cube.

First, for each pixel, the algorithm finds the minimum and maximum of its R, G, B values. Let C_max be the largest of R, G, and B and C_min be the smallest. The difference Δ = C_max − C_min is the range of the pixel's color.

The *hue (H)* component is determined by which color channel is dominant. If Δ = 0 (i.e. R=G=B, a gray pixel), the hue is undefined.

The *saturation (S)* component measures color purity. It produces a value between 0 and 1 (or 0–100%) indicating how vivid the color is.

The *value (V)* component is simply the maximum of the R, G, B channels. This corresponds to the brightness of the pixel (i.e. on a 0–255 scale if using 8-bit color channels).

After conversion, each pixel is represented in HSV. This color space makes it easier to isolate green pixels: we can simply specify a range of H values corresponding to "green" and look at those pixels' S and V. In this system, the software defines a hue range for green (for example, a range around 120° hue) and possibly a minimum value (brightness) threshold to disregard very dark regions. All pixels falling within the green hue range and above the brightness threshold are considered candidate weed pixels. Instead of using the hue directly for further processing, the system uses the saturation channel output for those candidate pixels. In other words, it creates a mask or image where each pixel's intensity is equal to its saturation if the pixel is in the green range (and meets brightness criteria), or zero if not. The rationale is that green weeds typically have a high saturation in the green hue range, whereas background soil or dry vegetation will not.

Convolution filtering (noise reduction)

Next, the algorithm performs a convolution on the resulting "green saturation" image to enhance contiguous green regions and suppress isolated noise. This acts as a form of morphological filtering. The convolution may use a 2D filter kernel (mask) that is applied across the image. In this invention, a 3×3 Gaussian-shaped kernel with a standard deviation $\sigma = 1$ is chosen. This small Gaussian mask effectively gives higher weight to a pixel's immediate neighbors (up, down, left, right) than to diagonal neighbors. Conceptually, the mask might look like a circularly symmetric bump, emphasizing the center pixel and its closest adjacent pixels. As the convolution runs over the image, each pixel's new value becomes a weighted sum of its neighborhood's saturation values.

The effect of this filtering is to accumulate saturation intensity in areas where multiple adjacent green pixels are present. A cluster of green pixels reinforces each other's values and produces a stronger response after convolution. In contrast, a single isolated green pixel (which could be noise or a tiny irrelevant spot) will be averaged out with its nongreen neighbors and yield a much lower value after convolution. This step, therefore, smooths the mask and eliminates salt-and-pepper noise, while slightly enlarging and highlighting the areas that are likely to be actual weeds (since real weeds will cover several pixels in the image). The choice of a Gaussian kernel ensures a smooth, spatially symmetric aggregation of neighboring pixel values. The inventors opted for this filter shape because it does not depend on a specific orientation and provides a gentle blending, as opposed to a harsh binary morphological operation. It is a computationally light operation suitable for a microcontroller.

Thresholding and binary decision

After convolution, the processed image is essentially a continuous-valued "heatmap" of green intensity, where higher values correspond to concentrated green regions. The next step is to apply a threshold to convert this into a binary image. Any pixel with a value above a chosen threshold is marked as "weed present" (binary 1), and those below become 0. Typically, the threshold is determined experimentally or via techniques like Otsu's method, aiming to separate background noise from true weed signals. The result is a binary mask highlighting only the significant green clusters and filtering out faint or spurious signals.

At this stage, the microcontroller evaluates the binary mask to decide if a weed has been detected in the frame. Rather than triggering on a single pixel, the system considers the aggregate size or count of green pixels. It counts how many pixels in the binary mask are flagged as green. If the count exceeds a predefined minimum threshold (representing, for instance, a minimal weed size or coverage area), the system concludes that a weed (or clump of weeds) is present. This double-threshold approach avoids false positives—a very small green blotch that might pass the color threshold will likely still be too small in area to trigger the system. Only a substantial green area will set it off. Once the decision logic confirms a weed detection, the microcontroller prepares to send the output signal.

Output signal generation (result)

If a weed has been detected as described above, the microcontroller issues a binary output signal to activate the weeding mechanism. This signal is sent over the RS-485 interface to the machine's power/actuator module. As noted, the signal is a short pulse, typically in the order of a few hundred milliseconds, to ensure the tool activates long enough to affect the weed but not much longer. The exact duration can be tuned depending on the tool's requirements (for instance, an electric discharge might need 300 ms to effectively kill the weed). After sending the pulse, the system resets the trigger and continues processing the next frames, maintaining a continuous scan for weeds as the machine moves.

It's important to highlight the real-time performance optimizations implemented in this workflow. The system uses a form of double-buffering and pipelining to achieve 10–15 FPS processing without interruption. While the microcontroller is busy processing one frame (color conversion, filtering, etc.), the camera and direct memory access (DMA) concurrently capture the next frame into an alternate memory buffer. The firmware toggles between two memory buffers: one is being processed while the other is being filled with new image data. Once processing on the current frame completes and a decision is made, the algorithm immediately begins processing the next frame that has just been captured in the other buffer. Meanwhile, the camera/DMA will start capturing the subsequent frame into the first buffer again. This ping-pong buffer technique ensures there's virtually no downtime between frames—the camera is always acquiring, and the processor is always either processing a frame or ready to start the next. It creates an assembly-line effect (a pipeline) between image acquisition and image analysis. Thanks to the relatively low complexity of each step (color conversion, a 3 × 3 convolution, and thresholding are all efficient operations), a mid-range microcontroller is sufficient for handling the workload in real time. Additionally, by using on-chip peripherals, such as DMA and DCMI, the system leverages hardware acceleration where possible.

Advantages over existing systems

The described sensing system offers several clear advantages and technical improvements over prior art and conventional methods:

- **Efficient onboard processing.** The image-processing method is streamlined and computationally efficient, avoiding heavy operations like large convolution kernels or cloud-based analysis. By using a simple color-space conversion and a small convolution mask, the system runs on a microcontroller instead of requiring a high-end CPU or GPU. This lowers the cost and power requirements significantly while still achieving effective weed detection.
- **Real-time detection.** With a processing rate of up to 10–15 FPS, the system can

operate in real time as the machinery moves through the field. This rapid response is crucial for practical weeding operations, ensuring that even at moderate speeds, weeds are detected and eliminated without delay. The use of double-buffering and DMA further optimizes throughput, something many older systems did not implement.

- **Reduced false positives via HSV filtering.** By focusing on the hue and saturation indicative of green plants and applying convolution to require clustered pixels, the algorithm dramatically reduces false triggers from noise or single leaves. Prior systems that lacked such filtering either had to tolerate more false positives or raise their detection thresholds (missing smaller weeds). This system's approach balances sensitivity and specificity, requiring a cluster of green for activation, which is more robust in field conditions (e.g. it will ignore a lone green pixel from a piece of debris or a distant crop row).
- **Low bandwidth and lightweight data handling.** Using the RGB565 format for camera output and converting to HSV on the fly means the data bandwidth and memory use are optimized. The system doesn't need to transmit large image files anywhere—all processing is local. Competing approaches that send images wirelessly (for cloud processing or remote monitoring) consume bandwidth and are vulnerable to signal loss. Here, only a tiny binary signal is output, and it's sent over a wired robust connection.
- **Robust communication and integration.** The direct wired connection (RS-485) between the sensor module and the weeding machine's actuators eliminates the radio communication failures and interference issues that were present in some prior art systems. RS-485 also allows the sensor to be physically distant from the main controller (e.g. at the end of a tractor boom or on a drone tether) while still maintaining a reliable link. This improves operational reliability in noisy farm environments. Moreover, the sensor is directly integrated with the machine's power module, simplifying the overall system architecture and response time.
- **Cost-effectiveness.** Because the system can run on an inexpensive microcontroller and does not require high-end processing units or extensive cloud services, it can be produced at lower cost. There is no need for large memory or storage on the device (each frame is processed on the fly and discarded unless a weed trigger event is recorded). This makes the solution more accessible for widespread deployment. Competing solutions that used advanced processors or heavy computing had higher costs that hindered adoption. By contrast, this invention achieves the necessary functionality with much simpler hardware, thereby lowering the barrier to adding it on farm equipment.
- **Versatility and safety.** The system's output is a generic trigger signal, which means it can be paired with various weeding mechanisms (electric shock, mechanical remover, targeted sprayer, etc.). It provides the detection component that can enhance many types of weed control machines. Additionally, by precisely targeting only green weeds, it can reduce the usage of herbicides (in sprayer setups) or ensure electrical or mechanical methods are applied only where needed, improving safety and reducing collateral damage to crops or soil.

In essence, the invention improves weed sensing by making it simpler, faster, and more reliable. It resolves the high processing requirements and communication bottlenecks seen in earlier systems with a clever combination of a focused color-space algorithm and integration of the sensor and actuator.

Conclusion

The system is grounded in a clear technical strategy: leverage the distinctive green color of most weeds to detect them with minimal computation. By using an onboard CCD camera and microcontroller, the design brings the "eyes" and "brain" together in one unit attached to the farming equipment, avoiding the pitfalls of remote processing. The technical description detailed how raw RGB images are converted to the HSV color space to isolate green hues, how a convolution filter accentuates genuine weed clusters, and how thresholding yields a decisive trigger signal. The hardware architecture—including the camera interface, power conditioning, and RS-485 communication—ensures that this processing translates into timely action by the weeding tool.

Crucially, the system achieves these tasks at 10–15 FPS using inexpensive hardware, demonstrating that complex weed detection can be made efficient. The double-buffered pipeline and use of DMA highlight the practical engineering techniques employed to meet real-time requirements. In field operation, the system would continuously scan the ground, and each time a weed is detected, it would promptly activate the weeder to eliminate the plant, all within a fraction of a second.

By addressing the limitations of prior art (complex algorithms, unreliable communication, and high costs), this invasive plant sensing system offers a robust and cost-effective solution for precision agriculture. It allows farmers and agricultural machinery to respond instantly to weed presence, reducing the need for blanket herbicide applications and enabling more eco-friendly weed management. In conclusion, the described system represents a significant step forward in automating weed control—combining imaging technology with smart processing to differentiate weeds from crops in real time and initiate targeted action. This integration of sensing and actuation in a compact, efficient package can greatly enhance the effectiveness of weed management in modern farming, leading to higher crop yields and lower environmental impact.

Embedded Image Sensing for Planting Line Identification in Precision Agriculture

The proposed system refers to PCT/BR2025/050051, filed on February 11, 2025. It is based on the previous "Green-on-brown sensing system for identifying invasive plants" described in PCT/BR2024/050019, filed on November 21, 2024.

Introduction

In the evolving landscape of precision agriculture, the need for accurate, real-time perception of crop environments has never been more essential. Among the most critical tasks in this domain is the identification of planting lines—the deliberate and methodical positioning of crops in rows that dictate every subsequent action, from fertilization to mechanical weeding. While GNSS-based guidance systems and advanced machine vision technologies have offered partial solutions, the complexity of field environments, combined with the cost and technical demands of high-end systems, have left a gap in the market for robust, affordable, and low-complexity solutions that can be deployed directly on agricultural machines.

This invention explores a novel approach to this challenge, based on a system of image acquisition and processing that performs all computations locally—in the field, within a compact embedded microcontroller. The system described was developed to identify planting lines through the detection of green vegetation patterns in real time, using a series of processing steps that emphasize minimal computational overhead while maintaining effectiveness and speed. It is particularly suited for integration with machines designed for nonchemical weed control, such as electrical weeding equipment, where spatial precision is a prerequisite for efficacy and safety.

Technical background and system overview

The green-on-brown sensing system described in the previous section applies a binary threshold. In this development, each pixel is evaluated against a preset value of green intensity or concentration. If it exceeds the threshold, it is marked as "active" or "plant-present"; if not, it is treated as background. The resulting binary image is essentially a map of where the system believes vegetation is located. These binary images form the basis for geometric analysis.

At this point, the system does not yet know whether the detected green regions correspond to a planting line or to random vegetation, such as weeds or outliers. To determine this, it performs a spatial analysis of the detected regions. After a binary thresholding operation has reduced the camera image to a high-contrast map of plant-like regions, the next step involves interpreting these regions not just as isolated patches of green, but as structured spatial formations that reveal intentional human planting patterns—specifically, crop rows. The idea is to move from raw detection of "green presence" to a more semantic understanding of spatial arrangement: a planting line.

Line identification

To identify lines, the system employs a set of simplified yet effective geometric analysis techniques that have been selected and adapted specifically for real-time performance on embedded microcontrollers. One of the foundational methods is the Hough line transform, a mathematical technique used in computer vision to detect lines in a binary image. However, the traditional implementation of this algorithm is computationally expensive and unsuitable for lightweight microcontrollers. Therefore, in this system, a simplified version of the Hough transform is used. This version restricts angle resolution and distance granularity to reduce computational load. Instead of searching across all possible line orientations, it focuses on detecting lines near the vertical or slightly angled positions, which corresponds to the expected layout of planting rows in most agricultural fields. This is a critical assumption that allows the algorithm to discard large portions of the search space, significantly accelerating processing time without major loss in accuracy.

Another powerful method included in the pipeline is sliding window histogram clustering. In this approach, the image is divided into narrow vertical slices (columns), and within each slice, the number of "active" green pixels is counted to generate a histogram of vertical density. These vertical slices are then scanned horizontally across the image using a sliding window to detect contiguous areas with high green density. If multiple neighboring windows exhibit similar histogram profiles, they are grouped into a cluster and evaluated as a potential row. This technique is especially effective when some parts of the row are obscured—by weeds, shadows, or plant variability—as it leverages local density peaks rather than requiring continuous visibility.

In tandem with histogram clustering, the system also performs a columnar density analysis, which provides a coarser but faster mechanism for identifying planting lines. In this method, each column of the binary image is evaluated for the number of active pixels, and the result is plotted as a vertical bar graph. Peaks in this columnar distribution represent areas where green pixels are most concentrated, which tend to align with the location of planting lines. This approach is particularly efficient when rows are nearly vertical in the image frame, which is often the case due to the mounting angle of cameras on agricultural machinery.

Together, these three methods—Hough transform, sliding window histogram clustering, and columnar density analysis—work in concert to extract geometric patterns from what begins as raw pixel data. The diversity of techniques ensures robustness: when one method fails due to occlusion, lighting, or irregularity, another can still yield a positive detection. For instance, if the planting line is curved due to terrain undulation or mechanical deviation during seeding, the Hough transform may struggle, but sliding window histogram clustering can still aggregate local density pockets and reconstruct the probable line orientation. This redundancy ensures that the system maintains high reliability under real-world conditions (Fig. 10.1, Fig. 10.2, Fig. 10.3).

Moreover, these methods include mechanisms to track small discontinuities in the row—gaps caused by failed germination, pest damage, or recent harvesting. Rather than treating these as termination points, the algorithms interpolate the missing data and maintain line continuity if the surrounding patterns strongly suggest a continuation. In practical terms, this means that the machine will not mistakenly deactivate its weeding tool just because a short section of a row is temporarily empty. This ability to infer and reconstruct planting lines despite missing data is essential for effective operation in live agricultural settings.

Once the presence of a planting line is confirmed, the system sends a signal to the main controller of the agricultural machine. This can be done via the RS 485 protocol—a widely used industrial communication bus known for its robustness in electrically noisy environments. The message is binary and time-bound, indicating that a planting line has been detected and specifying the duration of action. In practical terms,

Fig. 10.1. Artificial intelligence (AI)-created image of a crop.

Fig. 10.2. Identification of line and not in line (targeted) plants.

if the system is coupled to an electrical weeder, this signal would activate the high-voltage electrodes for a brief, predefined time—i.e. around 300 ms—ensuring that only the region overlying the detected planting line is treated.

This design eliminates the need for high-bandwidth wireless communication or cloud-based analysis, both of which are impractical in remote agricultural fields. It also reduces energy consumption, because the microcontroller only activates the weeding apparatus when a target is positively identified. This not only saves battery power but also reduces unnecessary wear on the mechanical components of the machine.

One of the core innovations of the system is its ability to perform all of this on a low-power embedded microcontroller. This contrasts sharply with other approaches found in prior art, where image data is uploaded to a remote server for analysis using AI or where real-time graphics processing unit (GPU)-based CNNs are employed. Those systems, while powerful, require extensive hardware, high energy budgets,

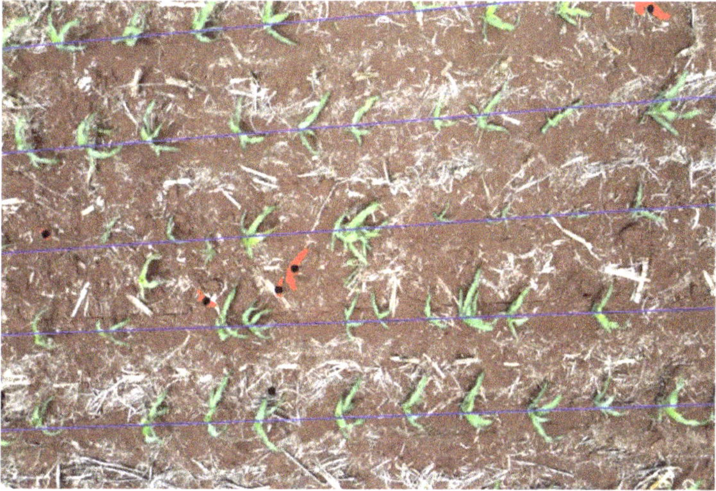

Fig. 10.3. Application in a real case.

and continuous connectivity—constraints that make them unsuitable for many agricultural use cases.

Beyond the technical architecture, the implications for the field are considerable. The system offers a viable pathway toward fully autonomous, closed-loop crop management systems. It can be deployed not only on electrical weeders but also on robotic tractors, targeted fertilizer dispensers, or spray systems for localized pesticide application. Its modular nature means that it can be embedded in handheld equipment, towed machines, or even integrated into aerial platforms such as drones with minimal adaptation.

In practical terms, the use of this technology could significantly improve the efficiency and sustainability of agricultural operations. By enabling real-time, high-precision actions based on actual plant presence rather than fixed GNSS coordinates or preprogrammed maps, farmers can reduce the use of herbicides, conserve energy, and minimize collateral damage to crops. This is particularly valuable in organic farming, where nonchemical interventions are preferred, and in emerging economies, where low-cost automation could greatly enhance productivity without increasing environmental impact.

In conclusion, the planting line identification system described in this section represents a shift in the design philosophy of agricultural sensing tools. It favors simplicity over complexity, reliability over novelty, and functional integration over technical abstraction. By focusing on what is essential—detecting green, linear regions with minimal delay and computational effort—it opens the door to practical, scalable, and sustainable precision agriculture technologies.

Confidence thresholding and debounce

Despite the power of the geometric analysis layer, the agricultural field presents a uniquely challenging environment for machine vision. Wind movement, animal interference, light flicker from clouds, shadows from equipment, and even overlapping foliage can introduce momentary anomalies in the binary image that could confuse the system. To address this, a final verification stage is introduced, known as confidence thresholding and debounce filtering. This layer acts as a cognitive buffer between raw detection and mechanical action, ensuring that decisions made by the machine are not only accurate but also stable.

The first line of defense in this layer is the use of temporal consistency checks. Each potential planting line detected in a frame is assigned a confidence score based on its similarity to lines detected in previous frames. If a line appears in a single frame but not in those preceding or following it, the system considers it likely to be noise or a false positive. Only if a line is consistently detected over a configurable number of frames—typically between three and five—is it accepted as a valid, actionable feature. This prevents the system from responding to one-off glitches in the data and ensures that machine actions are grounded in persistent environmental cues.

Building on this, the system employs a majority filtering mechanism, which essentially implements a voting logic across time. Each potential planting line is assigned a binary state (present or absent) in each frame. Over a sliding time window, the system calculates whether the majority of those frames agree on the line's presence. If a majority consensus is reached, the line is marked as "confirmed." If not, it is discarded as uncertain. This type of temporal filtering mirrors techniques used in digital signal processing to suppress jitter and prevent erratic behavior in control systems.

Another key benefit of the debounce mechanism is its ability to prevent rapid on–off toggling of machine components. Without debouncing, a partially obscured line that blinks in and out of visibility could cause the weeding electrode to activate and deactivate erratically—an outcome that is not only inefficient but also potentially damaging to crops and equipment. By enforcing a minimum duration for activation and deactivation events, the debounce filter ensures smooth transitions and machine stability.

This debounce logic also accommodates intentional lag, which is useful in certain field scenarios. For example, when the machine exits a section of the field with well-defined rows and enters an area with sparser planting, the system will maintain the last known valid row identification for a short time, anticipating re-emergence. If no further confirmation appears, the row is dropped. This feature acts as a cognitive "short-term memory" for the machine, allowing it to cope with temporary loss of visibility while still behaving rationally.

In practice, the combination of temporal checks, majority filtering, and debounce thresholds provides a powerful safeguard against the unpredictable nature of field conditions. It ensures that downstream systems—whether actuators, applicators, or loggers—are driven not by every flicker of green in the image, but by a carefully distilled signal representing the stable, consistent presence of a planting line. This design philosophy, blending geometric intelligence with temporal logic, is central to the system's overall robustness, and is a major reason why it succeeds in agricultural environments where more complex AI-driven solutions may struggle due to lack of reliability or overfitting.

Conclusion

The identification of planting lines using embedded image processing represents a pivotal advancement in the field of precision agriculture. By shifting the focus away from heavy computational models and instead embracing streamlined algorithms designed for real-world field deployment, this system achieves a rare combination of efficiency, reliability, and accessibility. The use of simplified geometric pattern recognition methods such as columnar density analysis, histogram-based sliding windows, and stripped-down Hough transforms enables the device to discern structured planting formations with impressive accuracy—even amid the visual clutter and inconsistencies common in natural environments.

Furthermore, the incorporation of temporal filtering techniques—namely confidence thresholding and debounce logic—adds a critical layer of decision stability that prevents false positives from influencing equipment behavior. This not only enhances the safety and reliability of downstream mechanical systems like electrical weeders but also improves the overall efficiency of agricultural operations by reducing unnecessary or mistimed interventions.

What sets this system apart is its foundational design philosophy: to work not under ideal laboratory conditions but in the dust, heat, and unpredictability of real fields. It balances technical rigor with engineering pragmatism, ensuring that the planting line detection operates with a fidelity sufficient for task execution, yet simple enough to be mass-produced and widely adopted by farms of all scales.

Ultimately, this approach demonstrates how thoughtful integration of computer vision, embedded processing, and agricultural context can produce a system that is not only technologically sound but also economically and operationally viable. In doing so, it paves the way for broader adoption of intelligent automation in agriculture, empowering farmers with tools that are both smart and grounded in reality. As agriculture continues to evolve toward sustainability and precision, systems like these will play a critical role in bridging the gap between innovation and implementation.

The Tethered Drone Pivot System

In the age of smart agriculture, the convergence of drone technology, computer vision, and high-voltage electrical systems offers a revolutionary approach to weed management. This section presents a groundbreaking invention: an electrical weeding system that utilizes a tethered drone mounted on or coordinated with agricultural machinery—pivots, tractors, or planters—to detect and eliminate undesired plants with surgical precision. This innovation epitomizes the fusion of sustainability, efficiency, and automation (Coutinho Filho and Rinzler, 2022).

Weed control has long been a challenge in agricultural practices. Traditional methods such as manual removal, chemical herbicide application, and mechanical disruption are increasingly questioned for their inefficiency, environmental impact, and the extensive labor they demand. The introduction of electrical weeding technologies, which destroy weeds by delivering controlled pulses of high voltage, provides a compelling alternative. However, existing systems often struggle with mobility, energy delivery, precision, and integration into standard farming routines.

The new solution described in this section addresses these limitations through the implementation of a tethered drone-based electrical weeding system. This mobile, intelligent, and scalable device is designed to integrate seamlessly with established agricultural infrastructure, transforming how farmers manage invasive plants in their fields.

Prior inventions and patents have laid foundational groundwork for automating weed treatment, including robotic arms with vision-based targeting, automated spraying systems, and electromechanical applicators that utilize mapped plant positions. Yet, these systems frequently encounter barriers such as high computational costs, limited signal reliability in rural zones, and financial infeasibility for widespread use. Systems described in previous patents introduced vision-based robotic weeding but could not fully overcome the challenges of integration with agricultural machinery and reliable power supply over extended periods.

The newly developed weeding system is structured into three primary layers—physical, logic, and application—each fulfilling critical and complementary roles. The physical layer includes the tangible components responsible for performing tasks in the field. This includes a drone tethered to a power source and mounted on a mobile or fixed carrier such as a tractor or irrigation pivot. Attached to the drone is a vision sensor that captures real-time images of the soil and plant life. The system includes an electrode designed to contact the target weed and deliver a lethal electric pulse, as well as a ground-based power module that supplies electricity safely and continuously. The power is delivered via a retractable tether, which maintains drone mobility while eliminating the need for frequent recharging. The choice of tether cable is crucial; it must support the necessary current while being lightweight and flexible enough to allow unhindered aerial maneuvering.

The logic layer is the intelligent core of the system. It contains the processing units and software that interpret sensor data, control flight and treatment actions, and store vital operational information. At the heart of this layer is the onboard computer and microcontroller system, which runs vision algorithms to detect green regions and crop lines. These algorithms allow the system to distinguish between planted crops and invasive weeds. The layer also features a georeferenced database, which stores detailed information about weed locations, environmental conditions, and historical flight data. Communication between the controller and onboard systems ensures synchronized and efficient drone operations.

The application layer serves as the human interface. It provides mission planning tools that define the drone's paths and treatment zones. The user can plan operations in advance, monitor the system in real time, and analyze flight records afterward to refine future missions. Operators retain the ability to start, pause, or stop the drone's operation at any moment. This upper layer ensures that the system remains accessible, safe, and adaptable for the end-user—typically the farm manager or machine operator.

Two primary embodiments of the system have been developed to address different agricultural needs. In one configuration, the drone is tethered to a central irrigation pivot (Fig. 10.4). The pivot provides both structural support and a convenient power and control hub. The drone moves along the pivot's path, inspecting and treating weeds beneath the irrigated zone without interfering with the irrigation process. This setup ensures consistent coverage of circular or arched crop zones.

In the second embodiment, the drone is mounted on a moving vehicle, such as a tractor or planter. In this scenario, the drone operates in parallel with the planting process. As seeds are deposited into the soil, the drone simultaneously detects and treats invasive plants, providing an efficient and streamlined field management routine. This design not only boosts productivity but also reduces the overlap of field operations.

The operation of the weeding process begins with the drone navigating its assigned route. The vision sensor captures high-resolution images of the terrain. These images are processed in real time to identify undesired plant features based on color segmentation algorithms, typically using HSV color space to isolate the green component associated with plant matter. Upon detection of an unwanted plant, the system positions the drone directly above the target. The electrode, carried by the drone, descends or is positioned to make contact. A precisely calibrated electric current is delivered, destroying the plant at its roots. Because the high-voltage converter is located on the ground or the carrier, the drone remains lightweight and maneuverable, enabling longer missions and reducing energy waste.

The scientific basis for this system's success lies in its ability to distinguish plants accurately and apply treatment efficiently. Advanced algorithms using green-on-brown segmentation, edge detection, and line mapping based on Hough and Radon transforms allow the system to identify weed regions and crop lines with high precision. This level of detail ensures that cultivated plants are not harmed during the process. The modular design also allows adaptation to various field conditions and crop arrangements.

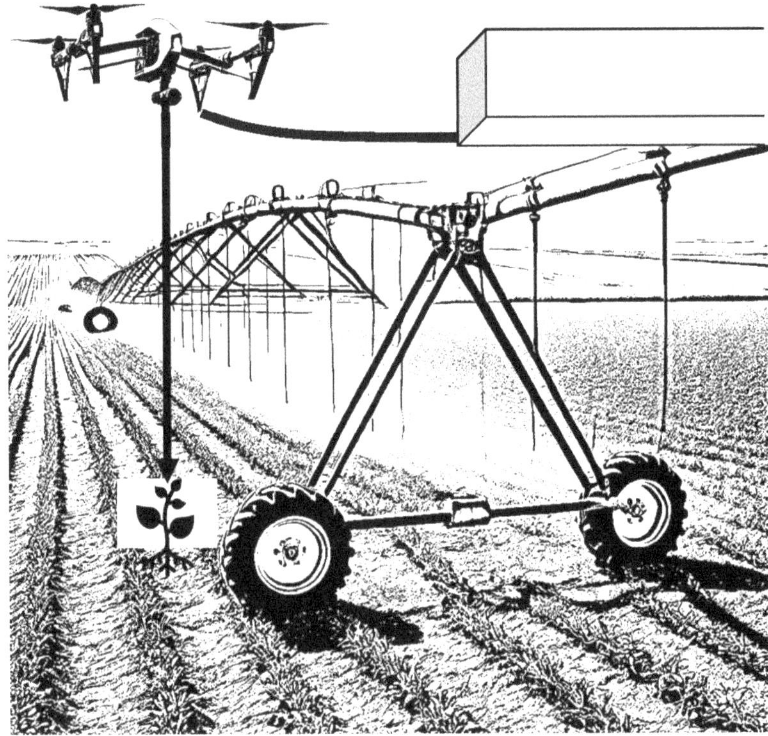

Fig. 10.4. Electrical weeding tethered drone connected to a pivot.

The use of tethered drones offers several key advantages over battery-powered aerial systems. Most significantly, tethered drones are not limited by battery life and can operate continuously for extended periods. The externalized power supply also enhances safety, as high-voltage systems remain grounded and away from delicate drone components. The tether ensures consistent electrical connectivity, reduces downtime, and simplifies maintenance operations.

From an environmental and economic standpoint, the system dramatically reduces reliance on chemical herbicides. By applying treatment only where needed, it minimizes the impact on nontarget organisms and reduces chemical runoff into soil and water systems. Precision targeting increases treatment efficacy while reducing labor and input costs. The integration of the drone system with standard agricultural equipment further lowers the barriers to adoption, making the solution accessible to small- and large-scale farmers alike.

In broader terms, the introduction of this weeding system paves the way for more comprehensive and autonomous crop management technologies. Future versions could incorporate disease detection, yield prediction, or real-time environmental analytics, contributing to the vision of a fully automated and sustainable farm.

In conclusion, the combination of electrical weeding and tethered drone mobility represents a significant advancement in the field of precision agriculture. This innovation integrates robust engineering, advanced software, and existing farm practices into a unified system that is efficient, sustainable, and ready for large-scale deployment. It exemplifies the shift toward smarter, cleaner, and more productive farming solutions suited for the agricultural challenges of the 21st century.

Drone-routing algorithms

Figure 10.5 compares the execution time of two algorithms—Christofides and Genetic—used for solving routing problems, particularly in the context of drone path planning. The X axis represents the size of the problem, which likely corresponds to the number of locations the drone must visit. The Y axis shows the average execution time required by each algorithm to compute a route. The data clearly demonstrates that Christofides consistently requires less execution time than the Genetic algorithm as the problem size increases.

In the field of drone operations, efficient routing is crucial. Drones are frequently tasked with covering a series of points for activities like crop monitoring, weed elimination, surveying, or delivery. This challenge maps closely to the classical traveling salesman problem (TSP), where the goal is to find the shortest possible path that visits each point exactly once and returns to the starting location (Fig. 10.6).

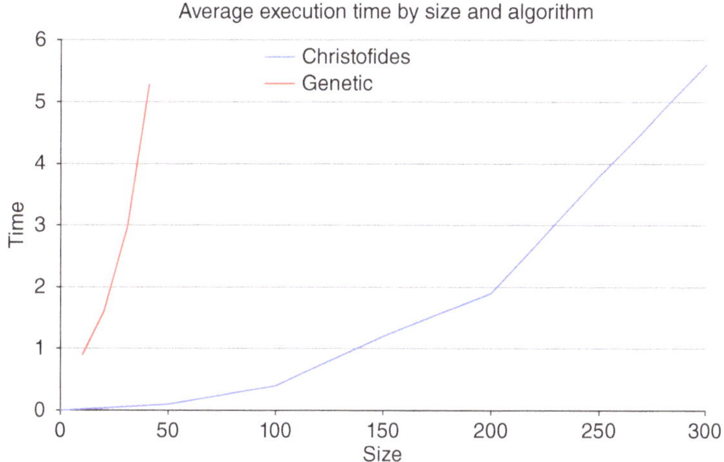

Fig. 10.5. Average execution time by size and algorithm type example.

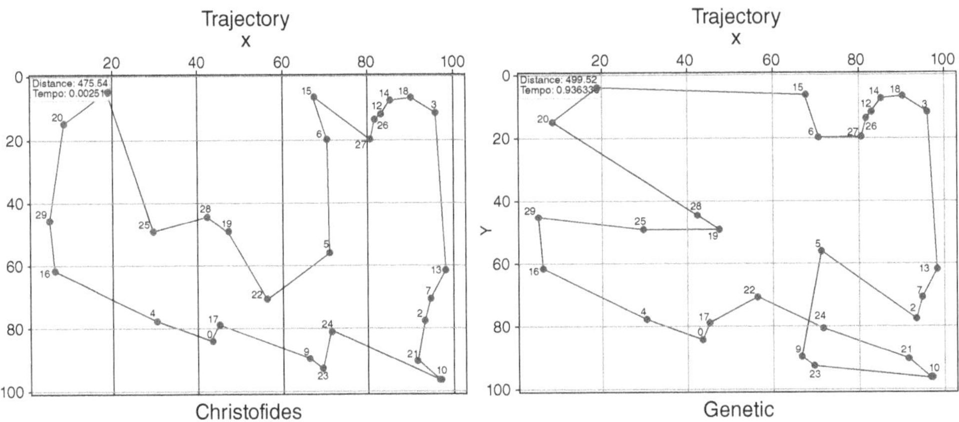

Fig. 10.6. Cases of travel map as a result of different methods.

The Christofides algorithm is a deterministic approximation method specifically designed for the metric TSP. It is recognized for producing a solution within 1.5 times the length of the optimal path. The algorithm proceeds through a series of structured steps: it begins by computing a minimum spanning tree of the graph, then it finds a minimum weight perfect matching among the vertices that have an odd degree. By combining the spanning tree and matching, a multigraph is created in which all vertices have even degree. From this multigraph, a Eulerian circuit is derived, and finally a Hamiltonian cycle is produced by shortcutting repeated vertices, preserving their order of appearance. This process results in a near-optimal route that is both efficient to compute and very close to the shortest possible path.

Christofides is especially advantageous when the goal is to minimize path length in a reliable and predictable way. It is particularly well-suited to environments where constraints are minimal, and the terrain is predictable. Its deterministic nature ensures consistent output, and its polynomial time complexity means that it remains efficient even for large problem sizes. However, it lacks the flexibility to handle additional constraints such as time windows, variable drone speeds, or dynamically changing targets. It is also less adaptable in scenarios that require on-the-fly re-planning.

The Genetic algorithm, by contrast, is a heuristic optimization method inspired by the principles of natural selection. It operates by maintaining a population of potential solutions, each representing a different path the drone could take. These candidate paths are evaluated using a fitness function, typically based on total distance or travel cost. The algorithm selects the most promising solutions, combines them through crossover operations, and introduces mutations to generate new candidates. Over many generations, the population evolves toward better solutions.

Genetic algorithms are highly flexible and can incorporate a wide range of real-world constraints. They can handle complex routing scenarios involving multiple drones, battery limitations, no-fly zones, and environmental data inputs. They are also well-suited for dynamic environments where targets or conditions may change mid-operation. However, this flexibility comes at a cost. Genetic algorithms tend to require significantly more time to reach a viable solution, especially as the problem size grows. Their stochastic nature means that results can vary between runs, and there is no guarantee of finding the optimal, or even near-optimal, solution unless considerable computational resources are invested.

In Fig 10.5, Christofides demonstrates a smooth, predictable increase in execution time as the input size grows, handling problems of size 300 with manageable time demands. The Genetic algorithm, however, shows a steep rise in execution time, even at small sizes. The absence of data points for the Genetic algorithm beyond a size of approximately 30 suggests that it quickly becomes impractical for large-scale applications due to excessive processing time or memory consumption.

The reason Christofides achieves shorter routes lies in its foundation in graph theory and its structured, efficient approach. By relying on the minimum spanning tree and perfect matching, it builds a foundation that closely mirrors the most efficient tour. The final steps in the algorithm ensure that the route is cleaned of redundancies, producing a streamlined path. Genetic algorithms, on the other hand, are susceptible to becoming trapped in local minimum, especially if the initial population lacks diversity or if the number of generations is insufficient for thorough exploration.

Choosing between these two algorithms depends heavily on the specific needs of the drone application. Christofides is ideal when route length must be minimized and when computational efficiency is a priority. It is especially effective for static environments with well-defined goals and no external constraints. Genetic algorithms are preferable when the routing problem involves complex variables and dynamic updates, and when the extra computation time can be justified by the need for adaptability and customization.

In conclusion, Christofides offers a mathematically grounded, efficient solution for computing short drone paths, while Genetic algorithms offer greater modeling flexibility at the expense of significantly increased computation time. Fig. 10.5 underscores this trade-off, highlighting the practicality of Christofides for large-scale routing and the need for caution when using Genetic algorithms for time-sensitive applications.

Soil Moisture-aware Control System: A Solution to How Different Electrical Resistances in Different Layers Impact Electrical Weeding

This is a general system, to be further explored throughout the section. Electrical weeding offers a sustainable, nonchemical approach to weed control, but its effectiveness is deeply influenced by the varying electrical resistivity of soil layers. This section introduces a soil moisture-aware control system designed to optimize electrical weeding by accounting for heterogeneous soil impedance. Drawing on Zasso's extensive simulations, field measurements, and system modeling, we explore the interaction between layered soil conductivity and electric field distribution, and how adaptive control strategies can enhance efficacy and safety.

The effectiveness of electric weeding depends not only on the electrical properties of the target plant but also on the medium through which current flows — predominantly, the soil. Variations in soil moisture, structure, and composition across different layers significantly impact current propagation and, therefore, the weed control outcome.

The resistivity profile of soil can change significantly over very short distances. The most critical factors influencing this vertical heterogeneity include:

- **Moisture gradients.** Soil water content is a dominant determinant of electrical conductivity. Water-rich layers conduct electricity more readily than dry layers, creating zones of differential resistance.
- **Organic matter concentration.** Generally, organic material has lower conductivity than mineral soil, especially when dry. Variations in decomposed plant material across soil horizons introduce significant resistive fluctuations.
- **Compaction and porosity.** More compacted soil layers often exhibit higher resistivity due to lower air permeability and restricted water flow. Conversely, loose, porous soil may retain more water, lowering resistance.
- **Electrolyte distribution (salts, pH).** Dissolved salts increase ionic conductivity. Soil layers with differing salt concentrations or pH levels can create impedance interfaces that distort the electric field.

Multilayer impedance modelling

To accurately simulate and control electric current flow in a real-world soil environment, impedance must be modeled across multiple discrete depth zones, represented as Zi (with i=1,2,...,ni=1,2,...,n), where each Zi corresponds to a soil layer of distinct electrical conductivity. Each layer is characterized by a specific conductivity, which denotes the layer index.

This stratified modeling approach enables:

- precise prediction of voltage drop across layers;
- analysis of current pathways and field strength distribution;
- estimation of energy delivered to plant root systems embedded at various depths;

As an example consider a layered soil system. Let:

- **Z1.** 0–5 cm—dry sandy loam, low conductivity;
- **Z2.** 5–10 cm—moist silty clay, moderate conductivity;
- **Z3.** 10–20 cm—saturated organic-rich soil, high conductivity.

In this case, the applied voltage will preferentially push current into the more conductive layer, influencing how the roots in it are exposed to lethal electrical fields—or in a practical scenario, this will determine how much energy is lost through the soil in each layer, once the current flows deeper into the roots searching (anthropomorphizing) for higher conductivities in the deeper soil layers.

In other words, the spatial location of weed root systems relative to conductive or resistive layers directly affects treatment efficacy. For example, a fibrous-rooted annual grass confined to Z1 may be underexposed if that layer is dry and resistive, while a tap-rooted perennial penetrating Z2 may receive more current and suffer greater electrocution damage.

Simulation insights on layered conductivity

To better understand how soil layering influences electrical treatment, a series of simulations were conducted using COMSOL Multiphysics. These simulations modeled voltage distribution and electric field strength under various soil configurations and conductivity profiles. Several key findings emerged.

When current is applied to dry or high-resistance top layers (e.g. dry sandy loam), voltage potential drops rapidly within the first few centimeters. This steep gradient significantly diminishes the electric field's penetration depth, reducing the likelihood of current reaching deeper root structures unless voltage levels are greatly increased. In such cases, efficacy is compromised unless adaptive voltage strategies are implemented.

Conversely, moist or saturated layers located deeper in the soil profile (e.g. organic-rich or silty clay zones) exhibit markedly higher conductivity. These layers form preferential pathways for current propagation, allowing electricity to more effectively reach deeper plant tissues. When root systems extend into or below these layers, the likelihood of full electrocution increases substantially.

Multilayer simulation models incorporating two or three strata with distinct conductivities demonstrate nonlinear behavior in step-voltage distribution. Unlike uniform soil models, voltage gradients do not diminish uniformly with distance from the electrodes. Instead, sharp potential differences are observed at the interfaces between layers of contrasting conductivity. This phenomenon is especially important for:

- **Safety analyses.** High voltage differentials can exist just centimeters apart, posing risks for nearby operators or equipment if not properly accounted for.
- **Efficacy targeting.** Understanding where voltage concentrates or disperses helps operators adjust equipment parameters for optimal weed control.

Together, these insights reinforce the need for real-time impedance sensing and adaptive control systems that account for soil variability in both design and operational phases.

Constructing a soil conductivity profiling sensor

Soil conductivity (or its inverse, resistivity) can be measured by applying a known voltage across electrodes inserted into the soil and recording the resulting current. By applying Ohm's law, the electrical resistance of the soil between electrodes can be calculated. When this process is repeated at various depths, a conductivity profile of the soil can be generated.

A practical soil conductivity profiling instrument would consist of the following elements:

- **Electrode probe rod.** A nonconductive rod with embedded conductive rings or pins placed at fixed intervals (e.g. every 5 cm). These act as input and sensing electrodes.
- **Multiplexer circuit.** A switching system that allows current to be applied between different electrode pairs. This enables selective measurement between shallow, intermediate, and deep layers.
- **Signal generator.** A low-voltage, alternating current (AC) source—typically, in the range of 1–10 kHz—to avoid polarization effects in wet soils and ensure stable readings.
- **Current and voltage Sensors.** Accurate measurement circuits to detect the current flowing and the voltage drop between electrodes.
- **Embedded controller or microcontroller.** A small computing unit (e.g. Arduino, Raspberry Pi, or industrial-grade microcontroller) to control switching, collect sensor data, and calculate conductivity.
- **Data logging and display interface.** Either onboard display or wireless transmission (e.g. via Bluetooth or LoRa) to visualize the data or export it to external systems.

Different embodiments may allow for dynamic readings to facilitate dynamic adjustments or mapping, such as probes that puncture the soil as equipment moves, or continuously with cutting disks at different distances or depths.

At minimum, the measurement procedure should have the following steps:

- **Insertion.** The probe is inserted vertically into the soil. It must maintain good physical contact with the surrounding soil to avoid false readings.
- **Sequential testing.** The system applies a small voltage between one electrode and another at a fixed spacing (e.g. 5 cm, 10 cm, 15 cm down). The resulting current is measured for each pair.
- **Calculation.** Using Ohm's law ($R = V/I$), and known geometry of the electrode spacing, the instrument calculates apparent resistivity and converts it to conductivity.
- **Profiling.** The controller assembles data from each depth and outputs a vertical conductivity profile, indicating zones of higher or lower electrical resistance.

Conclusion

The ability to measure and interpret soil conductivity across multiple depths is essential for optimizing electrical weeding operations. A well-designed profiling system enables operators to identify dry, resistive surface layers that may limit current penetration and require compensation. It also helps detect moist, conductive zones that naturally favor effective energy delivery to weed root systems.

Beyond immediate operational decisions, such instruments offer long-term benefits by allowing seasonal monitoring of soil changes, such as those caused by rainfall, irrigation, or agricultural cycles. When integrated with real-time control systems, this data supports adaptive voltage modulation—ensuring that the equipment delivers only as much energy as needed, tailored to the soil conditions at each moment.

Ultimately, coupling this profiling capability with autonomous or handheld applicators transforms electrical weeding into a smart, self-adjusting process. This not only improves treatment efficacy and consistency but also enhances safety and energy efficiency across a variety of agricultural and environmental settings.

Electrode Arrays with Dynamic Spacing Control for Power Stability

Introduction

Maintaining a stable electrical power output during soil-based electrocution operations is critical for operational reliability, equipment longevity, and consistent vegetation control performance. One of the principal variables influencing this power delivery is the soil impedance, which varies with moisture content, texture, compaction, and the mechanical coupling of the electrodes to the ground.

Traditional systems attempt to control power stability primarily by adjusting output voltage or using complex power electronics. However, a mechanical–electrical co-regulation approach—based on dynamically adjusting the spacing and pressure of electrode arrays—offers an alternative or complementary method, especially in environments where voltage regulation is limited, unsafe, or inefficient.

Description

As previously discussed, the electrical power (P) delivered into a resistive load such as the plant–soil system is defined by Ohm's Law. V is the applied voltage (often limited by hardware, safety, or regulatory constraints), and R is the load resistance, primarily a function of the soil and electrode configuration. Thus, to *stabilize power output* when V is fixed, bounded, or constrained into a band of values in the case of a self-adjusting system, the proposed solution must manipulate R mechanically, through adjusting the distance between the electrodes and, therefore, changing the soil resistance the system faces, which directly impact total resistance R. The key insight is electrode spacing and ground contact pressure directly influence the effective resistance of the soil path (R). There are a few alternatives to implement this.

By creating an actuated electrode spacing by increasing the distance between active electrodes, the effective resistance of the path through the soil increases due to longer current paths and higher energy dispersion. By reducing spacing, the current path shortens, and resistance drops—allowing more current flow at a given voltage. This creates a feedback mechanism where, when facing low resistance soil, the system spreads electrodes apart to raise R stabilizing P, and where, when facing high resistance soil, it brings electrodes closer to lower R stabilizing P.

Another alternative is creating an actuated height or pressure adjustment system at the electrode level, which may increase the band of the resistance faced. The band of change may be even higher if a flexible or spring-loaded electrode is used, as this adjustment may also change the area of contact. This can be used independently or in addition to an actuated electrode spacing. The electrode–soil contact quality affects impedance: resistance is raised by reduced pressure or contact area, especially in loose or dry soil; resistance is lowered by increased pressure or deeper penetration in compact or moist soil, or by enhancing the area of contact with either the soil or the plants the electrode touches. Thus, actuators can modulate ground pressure (either mechanically or via suspension elements) to add another degree of freedom in regulating R. Increasing pressure or decreasing height (increasing penetration) lowers electrical resistance, and decreasing pressure or increasing height (decreasing penetration) increases electrical resistance, therefore adjusting R to stabilize P.

Such a system may comprise a sensor array with real-time monitoring of delivered power, voltage, current, soil resistivity or contact pressure, or any combination of the above. A control algorithm calculates the required electrode distance and/or pressure to maintain a target power setpoint (P) and actuators to mechanically adjust either the electrode-to-electrode horizontal spacing, or the electrode height, and downward pressure or soil penetration (if spring-loaded or hinged). A feedback loop may ensure continuous adjustment for semiconstant power delivery, even as terrain and soil conditions change.

Conclusion

Dynamic electrode spacing and pressure modulation provide a mechanically elegant solution to stabilize electrical power delivery in soil-based electrocution systems. This approach is particularly useful when voltage cannot be freely modulated, offering a path to stable, energy-efficient, and consistent field performance across widely varying soil and environmental conditions.

Electrode Arrays with Dynamic Spacing Control to Ensure Penetration Depth

Introduction

Effective electrical weeding depends on the precise delivery of electric current deep enough into the soil to destroy not only the aerial parts of plants but also, critically, their root systems. Traditional methods using fixed electrode spacing face challenges because soil conditions, plant types, and moisture content vary across environments. A solution emerges by dynamically sensing the soil's electrical resistance—and possibly other variables—to intelligently adjust the distance between electrodes during operation. This ensures that the electrical current consistently reaches the necessary depth for root destruction.

Description

A system designed to achieve this goal revolves around three main functions:

- **Measurement.** A real-time sensor system measures the soil's electrical resistance (and potentially humidity, temperature, compactness, etc.).
- **Calculation.** A control unit calculates the optimum electrode spacing based on sensed data to achieve the correct penetration depth.
- **Adjustment.** Mechanized actuators dynamically move the electrodes closer or further apart during operation.

This ensures that energy is delivered efficiently and effectively, killing plants from leaf to root under all soil conditions.

Soil acts as a crucial part of the electrical circuit. Its electrical resistance determines how much current actually passes through the roots, how deep the current can penetrate, and how much energy is lost to the ground instead of affecting the plants.

Low soil resistance (e.g. in humid soils) may prevent current from reaching deep roots unless electrode distance or disposition, and applied voltage are optimized. On the other hand, dry and less conductive soils allow deeper and more efficient energy distribution but increase energy losses through the Joule effect in the soil.

Thus, the system must reduce electrode distance if soil impedance is high, concentrating energy between electrodes, and increase electrode distance if soil impedance is low.

Among the benefits:

- **Energy efficiency.** Reduces wasted electrical power by matching application intensity to real needs.
- **Improved efficacy.** Ensures roots are fully killed across all soil conditions.
- **Operational safety.** Reduces the risk of overvoltage or unintended arcs by matching the system's power levels to real-world conditions.
- **Sustainability.** Promotes better resource management, minimizes environmental impact, and enhances the viability of electrical weeding as an herbicide alternative.

Conclusion

Dynamic electrode distancing based on real-time soil resistance sensing represents a major step forward for precision electrical weeding. By adapting to environmental conditions on-the-fly, the system optimizes energy usage, maximizes efficacy, and ensures consistent root destruction. As agriculture moves toward greater sustainability and reduced chemical reliance, inventions like this provide a vital tool for the future.

Self-calibrating System for Different Weed Stages

The efficiency of electrical weeding depends not only on delivering sufficient energy to kill a plant but also on delivering the right amount of energy to the right kind of plant at the right time. One of the most significant challenges in achieving high systemic lethality is accounting for the variation in plant developmental stage, species type, and morphology. Traditional electrical weeding systems apply power uniformly, assuming that all plants present similar resistance profiles and vulnerability thresholds. In reality, however, weed populations are heterogeneous, composed of various species, growth stages, and root structures, each requiring a different electrocution strategy.

To overcome this challenge, a new approach has emerged: the self-calibrating electrical weeding system, driven by real-time image sensing and plant characterization. This section details how computer vision, species recognition, morphological inference, and machine learning converge to form a responsive system that dynamically adjusts its power delivery for maximum effectiveness and efficiency.

Introduction

The self-calibrating system is built around an integrated vision-based sensor array mounted on the weeding implement, whether handheld, vehicle-mounted, or autonomous. As the device moves across the field or target area, high-resolution images are captured and processed in real time. The processing pipeline applies trained CNNs and decision-tree logic to analyze key botanical features of each detected plant. From this analysis, the system identifies a combination of parameters that are crucial to determining electrocution strategy. These parameters include but are not limited to:

- **Overall plant size**, including both height and width, to estimate aboveground biomass and likely root system depth.
- **Leaf shape and structure**, used to distinguish between monocotyledons (monocots) and dicotyledons (dicots), which have differing vascular organizations and electrical conductivity profiles.
- **Species identification**, relying on databases that include known resistance levels and typical energy consumption for effective root-kill.
- **Morphology analysis**, estimating the aerial-to-root biomass ratio, which provides a clue about how deeply energy must be driven.
- **Weed stage classification**, determining if the plant is at an early seedling stage, vegetative growth, or reproductive maturity, all of which influence the required power.

From this set of data points, the control unit performs a power requirement calculation, adjusting the voltage and pulse characteristics delivered by the electrodes accordingly.

Technical mechanism

The core of the system is a high-speed embedded AI processor, typically powered by a GPU-enabled microcontroller or an edge tensor processing unit (TPU). This unit runs a plant classification algorithm that takes the visual input and classifies each plant instance within a defined field of view. The algorithm references a pretrained model, developed using thousands of labeled images from varied geographic regions and plant growth conditions.

After classification, the system makes an inference: based on prior empirical data, it maps each plant type and size combination to an optimal energy dose. For example, a young *Amaranthus* spp. (broadleaf dicot) may require a brief pulse at 3.5 kV to kill its root, while a mature *Cynodon dactylon* (a deep-rooted monocot) may need a sustained 8–10 kV application over a longer duty cycle.

Importantly, the system accounts not only for the species but also the individual condition of each weed. Two plants of the same species but at different stages of growth—or growing in different soil types—may receive different electrical treatments. The system performs this differentiation in milliseconds, enabling continuous calibration on-the-move.

A critical part of the mechanism is the feedback loop. The AI processor receives not only the image data but also electrical feedback from the electrodes—resistance, voltage, and current data during application. This feedback allows the system to learn in real time. If a power setting fails to kill a particular weed species effectively (e.g. the plant remains upright or does not show withering within a time frame), the system stores this instance and adjusts its parameters incrementally on future passes.

Morphological intelligence and electrocution optimization

One of the most innovative components of the system is its ability to interpret plant morphology as a proxy for electrical behavior. Morphology-driven parameters include:

- **stem thickness** as an indicator of internal moisture and vascular conductivity;
- **leaf-to-stem ratio** as a proxy for surface conductivity;
- **aerial biomass vs estimated root depth**, influencing the energy required to drive a current into the root crown.

This is especially useful when dealing with mature monocots such as *Sorghum halepense* or *Echinochloa crus-galli*, which possess extensive underground rhizome networks that require deep and sustained power delivery to ensure complete neutralization.

Conversely, small dicots, such as *Portulaca oleracea* or *Stellaria media*, have shallow roots and thinner cuticles, requiring much less energy. Applying too much power to these plants results in wasteful energy dissipation and increased risk of arcing or collateral soil impact. The self-calibrating system avoids such inefficiencies, ensuring that every microjoule is spent where it has the greatest physiological impact.

Species–specific calibration

In high-value crops or urban weed management applications, precision is paramount. The system maintains an internal lookup table associating each identified weed species with a lethal dose model. These models are built from field trials and lab measurements and include:

- minimum energy thresholds;
- time–voltage interaction profiles;
- resistance band data;
- optimal electrode polarity configurations.

When the vision system detects a known species, such as *Chenopodium album*, it immediately retrieves the associated parameters and tailors the electrocution sequence accordingly. If the species is unknown but morphologically similar to a known type, it defaults to a probabilistically safe configuration, erring on the side of efficacy without excessive energy waste.

Dynamic power delivery and electrode activation

The electrical subsystem integrates tightly with the sensing logic. Once the processing unit has determined the necessary energy level, the power electronic module governing the voltage source adjusts the duty cycle, voltage amplitude, and duration of application. In tandem, the electrode control system may reposition or reorient the electrode tips, depending on weed distribution and spacing, further increasing precision.

This synergy of sensor, processor, and electrode allows the system to execute *individualized treatments*, even within the same application row. For example, three weeds of different types and sizes, spaced only 20 cm apart, will receive entirely different energy profiles as the electrodes pass over them.

Data collection and machine learning loop

Another significant advancement is the system's learning capability. Each treatment is logged in a local database, along with pretreatment identification and posttreatment results (either manually confirmed or inferred by subsequent imaging). Over time, the system builds a localized treatment map, improving its accuracy for that specific farm, field, or environment.

This database enables the system to perform localized calibration without human intervention. In future seasons or in the next row of treatment, it applies a more informed power delivery strategy based on the unique resistance behavior of local soil, climate, and weed ecotypes.

By treating only what is necessary with exactly the required energy, the system reduces:

- energy consumption per hectare;
- equipment wear and tear;
- risk of electrical arcs and fire;
- collateral damage to nearby crops or soil biota.

This precision agriculture philosophy makes the technology suitable for both organic and regenerative systems where soil microbiome preservation is as important as weed suppression.

Conclusion

The self-calibrating system for electrical weeding, driven by real-time image sensing and adaptive energy logic, represents a major leap in agricultural precision. It merges deep learning with electromechanical control to create an intelligent, field-aware solution. Instead of treating the field as a homogeneous challenge, it treats each plant as a unique target, adjusting power delivery based on its biology, morphology, and resistance.

This level of dynamic personalization transforms electrical weeding into a high-resolution agronomic tool, capable of not just killing weeds, but doing so with scientific finesse and sustainable responsibility.

Field Potential Monitoring and Step-voltage Mapping System

Introduction

This section introduces a high-precision safety mechanism for high-voltage agricultural systems based on dynamic voltmeter architecture. The system operates through direct-contact electrode arrays embedded into the soil, which provide continuous real-time data on local potential gradients. These measurements are processed to compute the step voltage for a defined human stride of 1 m, considering a representative human body impedance of 1–2 kΩ. The calculated step voltage is used to determine, update, and enforce the minimal safety distance from the active high-voltage electrodes during electrothermal weed control operations.

Description

The system architecture is centered on a specialized voltmeter array formed by precision voltage-sensing electrodes driven into the soil at specified depths and intervals, typically spaced every 10–50 cm radially from the energized applicator perimeter. The sensors are connected via high-impedance differential amplifier inputs to a multipoint data acquisition unit. Each electrode functions as a passive probe with a low-leakage resistive interface to avoid influencing the native potential distribution in the surrounding soil.

The voltmeter reads absolute ground-referenced potentials at each electrode position. These scalar potential values are timestamped and geolocated. A discrete computational module processes pairs of readings spaced at 1 m, or another distance (d), from which can be understood the voltage/distance, simulating a human stride. For each pair, the voltage difference ΔV is calculated, and using Ohm's Law with the body resistance (R_body), the equivalent current (I_step = ΔV/R_body) is derived. However, since the goal is safety assessment rather than physiological current modeling, the ΔV is used directly as the decision metric.

To ensure robustness, the system employs both spatial and temporal averaging. Step voltages are averaged over several measurement lines (radial, tangential, diagonal) and over a sliding window of acquisition cycles, typically 0.5–1.0 s. These averaged step voltages are then compared against a defined safety threshold, commonly 50–60 V DC per the International Electrotechnical Commission (IEC) and DIN VDE guidelines (e.g. DIN VDE 0100-701 and DIN VDE 0101), with a 10–15% design margin.

The voltmeter and logic system are encapsulated in a weather-sealed, shielded enclosure compliant with IP67, powered by a redundant 24 V DC rail with uninterruptible power supply (UPS) backup. Data processing is performed on a hardened real-time embedded controller (e.g. Cortex-M7 or equivalent), with interrupt-driven analog-to-digital converter (ADC) sampling and DMA-buffered communications to a supervisory safety programmable logic controller (PLC) or main control unit via a controller area network (CAN) bus or industrial ethernet (i.e. PROFINET or EtherCAT).

If any calculated step voltage exceeds the defined threshold, the controller flags the condition and initiates a three-tier response system: (i) visual and audible local alert at the equipment; (ii) logging and remote transmission of the event data packet (including timestamp, location, voltage profile, and response code); and (3) immediate deactivation of the high-voltage pulse train by gating off the primary inverter bridge or decoupling the high-voltage transformer via a solid-state relay matrix. This process occurs in less than 100 ms from the point of detection, ensuring compliance with functional safety requirements (SIL-2 or higher).

The electrode voltmeter array is also used to calculate the dynamically enforced safety perimeter. The spatial potential field data is fitted using a 2D interpolative mesh (e.g. inverse-distance weighting or Delaunay triangulation), and the radial distance from the applicator at which all interpolated 1-meter ΔV values fall below the safety threshold is computed. This distance defines the instantaneous safe boundary, which is continuously updated and can be graphically overlaid onto a global navigation satellite system (GNSS)-based geofence visualization on the operator's human–machine interface (HMI).

Calibration of the system is performed prior to each use, with the voltmeter auto-null procedure and environmental baseline correction using a known neutral potential probe placed beyond the maximum field influence range. Sensor drift compensation and soil impedance normalization factors are applied via lookup tables embedded in firmware, based on empirical measurements across standard soil types (e.g. loam, clay, sandy).

In environments with layered or heterogeneous soil conductivity profiles, the system compensates by analyzing temporal stability and signal skew across the array. Electrode pairs showing anomalous gradients inconsistent with expected exponential falloff are flagged, and their values weighted down or discarded from safety perimeter computation (Schematic 10.1).

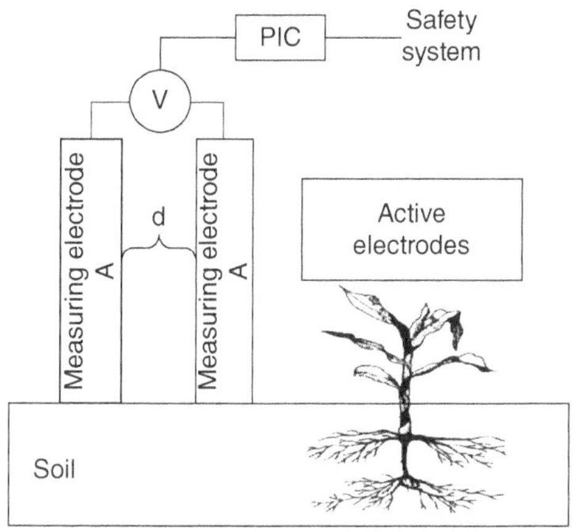

Schematic 10.1 Field potential monitoring and step-voltage mapping system.

This dynamic electrode voltmeter architecture allows for precision control and real-time assurance of safety distances in high-voltage field systems, particularly those operating with soil-distributed return paths. It eliminates the dependency on static simulations and worst-case assumptions, offering a context-sensitive, adaptive safety envelope responsive to environmental changes such as moisture, compaction, and salinity.

Conclusion

The system transforms passive soil voltage mapping into an active safety control mechanism, integrating real-world feedback into functional safety logic. It enables high-voltage weed control systems to maintain maximum operational efficiency while adhering to stringent human protection requirements in dynamically varying field conditions.

Real-time Root System Impedance Estimation

In the domain of electrical weeding technologies, the accurate estimation of plant impedance—especially that of the root system—is critical for optimizing energy delivery and ensuring effective plant control. Impedance is a dynamic property influenced by plant morphology, development stage, and root volume. This section presents an integrated approach combining real-time vision analysis and impedance modeling to estimate plant root system impedance and guide the electrical treatment parameters accordingly.

The root system of a plant, comprising fine roots, primary roots, and secondary structures, exhibits complex impedance behavior. This impedance is primarily a function of root architecture and volume—more extensive systems provide more conductive paths; moisture and electrolyte content—which influence the conductivity and soil characteristics, including depth-variable resistivity and heterogeneity of soil layers.

Impedance across a plant can be modeled as a combination of the aerial (air) system and multiple subsurface (root) paths. Each root path ($x_1 \ldots x_n$) is associated with a depth-dependent soil impedance ($z_1 \ldots z_n$), forming a parallel network through which current propagates.

The role of vision systems

Vision systems offer a powerful, noninvasive method to infer key morphological parameters of a plant that correlate with its electrical impedance. A real-time imaging system can classify plants based on species-specific morphology (e.g. monocotyledon vs dicotyledon), growth stage (size, volume, and height), and health indicators (e.g. color, rigidity, and transpiration patterns). These traits are essential in estimating how electrical current will behave as it moves through the plant's structure—from the aerial system, through the xylem and phloem conduits, into the roots, and finally into the soil.

One of the critical insights derived from visual analysis is the estimation of root volume, which directly impacts the energy required for effective control. Plants with larger, deeper, or more complex root systems provide greater overall conductive mass and often contain redundant vascular paths through which current can diffuse. As a result, more energy is needed to achieve lethal effects, such as electroporation or thermal collapse, across the full root network. If energy is insufficient, portions of the root system may remain viable, leading to plant regrowth. Vision systems trained to recognize indicators of extensive rooting—such as stem thickness, canopy width, or specific architectural features—can infer a lower root impedance but higher energy demand, prompting dynamic adjustment of power parameters to ensure complete root inactivation.

In parallel, the volume and architecture of the aerial system—leaves, stems, and branches—play a key role in shaping how the current initially enters the plant. A larger air system offers more surface area for electrode contact, effectively increasing the number of entry points for the current and lowering the local contact resistance. In this sense, the aerial biomass acts as a kind of "energy capture array," improving electrical coupling with the applicator. This has a dual effect: it reduces the initial impedance spike commonly observed at the onset of application and allows greater penetration of current into the vascular system, thereby enhancing treatment efficacy. High-definition imaging can detect large, healthy aerial systems and instruct the controller to leverage this increased conductivity by safely delivering more power with minimized loss at the interface.

Furthermore, visual indicators such as leaf orientation and rigidity provide clues about moisture content and internal pressure, which affect the conductivity of xylem and phloem pathways. For example, upright, turgid leaves imply well-hydrated vascular tissue and lower internal resistance—conditions that favor deeper current propagation. Conversely, drooping or desiccated tissues may require a different strategy, such as increased voltage or longer pulse duration, to overcome internal impedance. By integrating these visual metrics, machine-learning models can predict the complete electromechanical profile of the plant and optimize the weeding parameters in real time.

In summary, vision systems serve as the eyes of an intelligent electrocution platform, not only identifying what kind of plant is being treated, but also determining how much energy it can absorb, where that energy will flow, and how much is needed to ensure total and irreversible control. This synergy between visual inference and power management is key to delivering precision weed control in diverse and dynamic field conditions.

Vision-to-impedance estimation process

The system for estimating plant impedance in real time, using computer vision as the primary sensor input, is composed of a series of interconnected stages that translate visual plant characteristics into electrical behavior predictions. This integrated approach allows a high-voltage weed control system to adapt its output dynamically for each individual plant or group of plants in the treatment path.

The process begins with *image acquisition*, where a set of cameras—typically mounted on the weeding equipment in front of the applicators—capture high-resolution RGB and optionally multispectral images of the terrain. These cameras operate at frame rates sufficient to keep pace with the forward movement of the machinery (typically between 3–10 km h^{-1}), ensuring that each plant is captured and analyzed before it reaches the electrodes. The camera sensors are calibrated not only for color fidelity but also for depth approximation through stereo vision or structured light, allowing for a three-dimensional reconstruction of each plant's aboveground geometry.

Once the images are acquired, the data flows into the feature extraction stage, where CNNs or similar deep-learning models are used to isolate individual plants from the background. These networks are trained on large, annotated datasets containing various weed and crop species, under different lighting and field conditions. The neural architecture is designed to segment plant contours, compute leaf area indices, detect stem thickness, and estimate canopy depth and spread. These metrics serve as proxies for both the size and the hydration status of the aerial system, which in turn are predictive of electrical conductivity and energy requirements.

Following this, the system performs classification and regression using a dual-path neural framework. One path determines categorical characteristics such as species, family, and morphological type (e.g. dicot or monocot), while the second path uses regression techniques to predict continuous variables such as biomass volume, leaf area, and likely growth stage. The classification branch helps constrain the regression with species-specific parameters (e.g. a young maize plant will have different impedance behavior from a similarly sized dicot weed), thereby improving the accuracy of the overall model. These outputs are time-synchronized with GNSS or machine position data so that the system can correlate each prediction with a specific region of the ground or plant zone under treatment.

With these morphological parameters in hand, the system performs root volume inference by applying species- and stage-specific models that relate the visible aerial biomass to expected root architecture. These models are based on empirical agronomic data and lab-calibrated measurements that establish a statistically significant relationship between what can be seen and what lies underground. For example, if the vision system detects a plant with a thick stem, broad canopy, and upright turgid leaves, the algorithm may estimate a relatively large root mass with deep penetration, which implies low root impedance but high total energy capacity. This stage effectively treats the plant as an asymmetric biological conductor, where visible metrics help infer unseen impedance paths.

Next, in the impedance estimation stage, the inferred root volume, along with aerial contact area and expected tissue conductivity, is translated into a modeled resistance and capacitive profile for the entire plant system. This profile is expressed as an equivalent electrical circuit—often a ladder or branching network of resistors and capacitors—that simulates how current will flow from the electrode through the aerial parts, into the vascular conduits (primarily the xylem), and down to the roots where it exits into the soil. The model accounts for dynamic soil resistivity, temperature, and even moisture distribution if auxiliary sensors are integrated.

Finally, in the control adjustment stage, this estimated impedance profile is used to dynamically tune the output of the high-voltage power system. Parameters such as output voltage, pulse width, rise time, frequency, and waveform shape are modified in real time to match the plant's specific electrical load. This adjustment ensures that the energy delivered is sufficient to destroy the plant's physiological function (usually via irreversible electroporation, dielectric breakdown, or thermal collapse of tissues), while avoiding excessive energy dissipation in the soil or through arcing losses. This precision enables not only increased efficacy and energy efficiency but also enhances operator safety and protects nearby crops or structures.

The entire pipeline, from image acquisition to power adjustment, is designed to operate within a tight latency envelope—typically under 100 ms—ensuring responsiveness even at practical field speeds. This continuous loop of perception, inference, modeling, and control is a foundational component of the next generation of intelligent, data-driven electrical weed control (EWC) systems. It is further characterized by its ability to self-improve over time, as posttreatment plant responses (e.g. wilting, regrowth) are logged and fed back into the model for retraining, creating an evolving intelligence that mirrors the diversity and complexity of field conditions.

Conclusion

In conclusion, the integration of real-time vision systems with high-frequency EWC represents a transformative advance in precision agriculture. By correlating visual plant features—such as aerial morphology, growth stage, and health indicators—with subsurface impedance behavior, the system enables adaptive power delivery that is biologically informed and electrically efficient. This not only ensures that each plant receives the appropriate energy dose to guarantee effective inactivation, but also significantly reduces energy waste, minimizes collateral soil impact, and improves operator safety.

The vascular architecture of plants, particularly the xylem and phloem, plays a pivotal role in mediating electrical conductivity between the air and root systems. A robust aerial structure facilitates greater current capture through increased contact area and reduced initial resistance, while an extensive root network demands more energy to ensure complete system breakdown. Vision-based inference of these anatomical characteristics enables dynamic modeling of each plant's resistive circuit, allowing real-time adjustments of voltage, pulse frequency, and waveform. This biologically nuanced approach enhances treatment efficacy across species and environments, reinforcing the reliability and repeatability of electrical weeding as a practical alternative to chemical herbicides.

Ultimately, the vision-to-impedance estimation pipeline protects and advances Zasso's technological edge by embedding deep biological insight into power electronics control. It builds on the company's core intellectual property of impedance matching and intelligent energy application by fusing it with real-time plant diagnostics. As a platform, it sets the stage for even greater autonomy, enabling systems that not only recognize and eliminate weeds effectively but also learn and optimize continuously through field feedback. This convergence of sensing, modeling, and adaptive control solidifies electrical weeding as a foundational pillar of sustainable, nonchemical weed management in the 21st century.

Autonomous Electric Weeding Robot with Machine Vision

In the context of modern agriculture, the challenge of weed control has become increasingly complex due to environmental, regulatory, and economic pressures. Traditional chemical herbicides are facing mounting scrutiny, and many are being phased out or restricted in various jurisdictions. In response to this, a new wave of technological innovation is transforming how we manage unwanted vegetation. Among the most promising developments is the deployment of autonomous small robots that utilize high-voltage electrical current to eliminate weeds. These robots are guided by GNSS systems and enhanced with advanced machine vision technologies capable of green-on-brown segmentation, enabling them to identify and target individual weeds with high precision.

This idea explores the convergence of electrophysical weed control and robotic autonomy. It delves into the engineering principles behind electrical weeding, outlines the integration of visual recognition and spatial localization, and evaluates the efficacy and limitations of these systems in field applications.

At the core of an autonomous electrical weeding robot lies a high-voltage power unit, typically a modular converter-transformer assembly capable of delivering several kilovolts in pulsed or modulated direct current (DC) form. This system is responsible for generating the energy required to ensure lethal current delivery to the weeds. The electrodes, often constructed from stainless steel or similar conductive materials, are mounted to an adjustable frame and designed to make consistent contact with the target vegetation. They are arranged in pairs, with positive and negative terminals positioned to channel the current effectively through the weed.

The robot's propulsion and control can be managed by a battery-powered drivetrain and onboard microcontroller. Navigation can be achieved through real-time GNSS data, often enhanced with real-time kinematic (RTK) correction to provide centimeter-level accuracy. This precision may be essential to ensure the robot can operate between crop rows and navigate complex field geometries without damaging desirable vegetation.

To safeguard both the equipment and surrounding life forms, a suite of safety mechanisms is integrated. These include voltage clamping strategies, arc suppression circuits, and grounding fault detectors. For human safety, insulation materials, physical barriers, and emergency cutoff systems are essential, especially during maintenance or transport. The system may also require that operators wear specialized protective gear, such as electrostatic discharge (ESD)-compliant boots, and maintain designated safety distances from the machine during operation.

The robot's ability to discriminate between crops and weeds can be underpinned by a green-on-brown vision system. This typically involves a combination of RGB or multispectral cameras and machine-learning algorithms trained to recognize the spectral and morphological signatures of vegetation against the background of bare soil. Convolutional neural networks (CNNs) are frequently employed to segment the visual field into discrete plant and nonplant regions.

Once vegetation is identified, the software calculates the location and boundary of the plant. This information is then passed to the robot's motion control system, which adjusts the robot's position and aligns the electrodes with the target. Timing is critical, as the electrical pulse must be delivered with millimeter-level precision to ensure that the current path encompasses the plant's conductive tissues while minimizing collateral effects.

The image recognition software is often trained on a diverse set of crops and weed species to improve reliability under varying light, soil, and weather conditions. In real-world scenarios, the system must perform under complex environmental factors such as changing shadow patterns, plant overlap, and partial occlusions.

Navigation in autonomous weeding robots depends heavily on accurate geolocation and dynamic path planning. RTK-GNSS systems provide the level of positional fidelity required to operate safely in densely planted fields. Typically, the GNSS module is coupled with an inertial measurement unit, which provides additional orientation and movement data to smooth navigation in cases of temporary signal loss or terrain variation.

The robot's route may be calculated using algorithms capable of real-time path optimization. These include A* (A-Star) for obstacle avoidance and rapidly exploring random trees for adaptive pathfinding. Obstacle detection is achieved through a combination of light detection and ranging (LIDAR) sensors, ultrasonic range finders, and bump sensors, allowing the robot to identify and navigate around unforeseen obstructions such as rocks, fallen branches, or animals.

In operation, the robot moves along predefined paths, periodically adjusting its course to match GNSS waypoints and visual weed detections. The fusion of spatial and visual data ensures the applicators are only activated when a weed is detected and properly aligned, thereby conserving energy and avoiding damage to non-target areas.

The integration of electrical weeding into autonomous robotic platforms presents a transformative approach to sustainable weed management. By replacing chemical herbicides with targeted electrical pulses guided by GNSS and machine vision, this technology offers an environmentally sound and highly effective solution to one of agriculture's most persistent problems. While technical challenges remain, particularly in the areas of energy optimization and machine-learning robustness, the trajectory of development strongly supports the eventual widespread adoption of this method in both organic and conventional systems.

As research progresses, these robots are likely to evolve from niche tools into essential components of smart farms, operating continuously, precisely, and autonomously to maintain weed-free fields with minimal human intervention and no chemical input.

Multimaterial Composite Applicators

The development of materials for agricultural technology, particularly for high-voltage equipment such as electric weeders, requires a delicate balance between structural strength, electrical insulation, manufacturability, and weight. Traditional construction materials, such as steel and concrete, offer excellent mechanical performance, especially in resisting compressive and tensile forces. However, they fall short in several critical domains when applied to high-voltage systems. These include electrical conductivity, weight, corrosion potential, and manufacturability in low-volume or modular production.

To address these challenges, players in the electrical weeding industry (more specifically Zasso) introduced a new class of composite materials inspired by the steel-reinforced concrete model but engineered specifically for applications requiring excellent electrical insulation, mechanical robustness, and low weight. This system replaces steel with fiberglass rebars and conventional concrete with a polymer matrix filled with ceramic oxides such as alumina. By using lightweight fillers in non-load-bearing regions, the composite achieves an optimal balance of performance and mass, particularly well-suited for casing, supports, shields, and structural members in electric weeding systems.

In civil engineering, reinforced concrete combines the compressive strength of cementitious materials with the tensile resilience of steel rebar. This synergy allows structures to endure both axial and bending stresses. The material system presented in this section retains that structural logic but replaces the conventional components with materials that better meet the requirements of high-voltage agricultural machinery.

The tensile elements in this composite are fiberglass rebars, long-strand continuous fiber rods with high strength-to-weight ratios and excellent resistance to environmental degradation. These rods are embedded in a matrix formed from a two-part thermosetting epoxy resin, into which finely ground ceramic oxide fillers such as alumina (Al_2O_3), silica (SiO_2), or zirconia (ZrO_2) are suspended.

This matrix—here referred to as "polymeric concrete"—provides exceptional compressive strength, excellent dielectric properties, and thermal resistance. It adheres tightly to the fiberglass rebars, forming a unified composite capable of distributing and absorbing complex mechanical loads.

Material composition and structural design

The matrix of this advanced composite consists primarily of epoxy resin, a versatile polymer known for its adhesion, low shrinkage, and stable mechanical performance. Epoxy alone is relatively weak under compression, which is mitigated by the addition of ceramic fillers. These fillers—especially alumina—confer superior compressive resistance, hardness, and electrical insulation, crucial for safe performance in high-voltage environments.

The fiberglass rebars, commonly formed from E-glass or S-glass filaments bound with a polymeric sizing agent, provide tensile strength comparable to mild steel but at a fraction of the weight. Their low thermal conductivity and dielectric properties make them ideal for structures that must support or encase high-voltage conductors without risk of short circuit or arcing.

To further reduce mass, nonstructural regions of the part—such as the inner cores or recesses—are filled with lightweight inert materials. These include polymer foams, hollow glass microspheres, or honeycomb inserts that provide form and stability without significant weight or cost.

The final structure, therefore, mimics a reinforced concrete beam in its form and function, but is fabricated entirely from nonmetallic, electrically inert materials, yielding a solution ideal for electric weeding equipment.

Fabrication process and geometric flexibility

One of the principal advantages of this material system is its ease of fabrication. The composite can be manufactured using low-pressure casting, filament winding, or pultrusion, making it accessible for both small-scale workshops and industrial production lines.

The process begins by placing fiberglass rebars into a mold configured according to the desired geometry of the part. Careful attention is paid to the orientation of the rebars to maximize tensile resistance along anticipated stress axes. The mold is then filled with the epoxy-alumina slurry, a viscous mixture that flows around the reinforcement and cures in place. Regions requiring low weight may be filled in advance with foam inserts or premolded hollow structures.

Curing is carried out at room temperature or under mild thermal acceleration, after which the part can be demolded, machined, or post-treated for enhanced surface properties. The ability to cast complex geometries—including ribs, mounting flanges, and embedded guides—allows engineers to tailor the material to the unique demands of agricultural machinery.

Mechanical and electrical performance

Laboratory evaluations and simulated stress testing reveal that this composite system delivers compressive strengths exceeding 150 MPa when alumina content reaches 60% by volume. Even at lower filler ratios, the material can endure significant static and dynamic loads, suitable for mounting heavy actuators, shielding power electronics, or acting as the structural backbone of robotic platforms.

Tensile strength, governed primarily by the fiberglass reinforcement, approaches 1,000 MPa when using high-grade unidirectional E-glass rebars. Bonding between the epoxy matrix and the fiberglass is critical and is often enhanced with surface treatments or silane coupling agents.

From an electrical perspective, the system is outstanding. The combination of ceramic fillers and nonconductive fibers results in a dielectric strength in excess of 20 kV/mm, with negligible surface tracking or partial discharge under high-frequency electric fields. This property is especially valuable in the context of electric weeding, where high-voltage pulses are applied near sensitive control and drive systems.

Thermal insulation is also favorable, with conductivities as low as 0.25 W/mK, providing resistance to rapid heating or arcing. The matrix and reinforcement are inherently noncorrosive, allowing long-term outdoor exposure without degradation, a stark contrast to a traditional metallic chassis.

Electric weeding equipment demands structures that can withstand significant mechanical stress, repetitive vibration, and proximity to lethal voltages without failing. Components such as applicator supports, electrode arms, robotic enclosures, shielding panels, and articulating joints all benefit from this composite material.

By embedding fiberglass rebars along stress trajectories—such as the bending plane of an applicator arm—engineers can design lightweight structures that resist both bending and torsion. The electrically insulating matrix ensures that no part of the structure can serve as an unintended current path, eliminating the risk of stray arcing or operator shock.

Furthermore, the low weight of the composite allows smaller robots to cover larger areas with greater efficiency and longer battery life. Combined with modular mold designs, the material system supports rapid reconfiguration for different crop types, row spacings, or electrode designs.

In field conditions, these composites are notably resilient. Tests with electric weeders operating at up to 15,000 V and 60 A showed no measurable degradation in structural or electrical performance over thousands of cycles. Impact tests using dropped weights confirmed the material's ability to absorb sudden shocks without cracking or delaminating.

Larvae Electric Control

This section presents a comprehensive analysis of the experimental development, adaptation, and validation of Zasso's high-voltage power module, usually used for electrical weeding, known as "Fulgur" for biological vector control. Targeting the larvae and pupae of the mosquito, *Aedes aegypti*, in stagnant water environments, the project integrates field-optimized power electronics, mobile deployment design, and tailored high-voltage application techniques. The efficacy of this method as a nonchemical vector control strategy is evaluated, with implications for municipal health and environmental management systems.

Introduction

The proliferation of *Aedes aegypti*, the primary vector for dengue, Zika, chikungunya, and yellow fevers, remains a critical public health concern in tropical and subtropical regions. The mosquito's life cycle is highly dependent on stagnant water for breeding, with the aquatic phases—egg, larva, and pupa—being vulnerable to intervention. This study explores the application of controlled high-voltage discharge using a mobile power module—Fulgur—to disrupt this life cycle.

The work falls within the domain of applied research and development (R&D), seeking to assess the technical feasibility, safety, and field-deployment potential of high-voltage techniques for targeted vector control in urban and peri-urban environments.

Experimental system design

The Fulgur power module was mechanically and electrically adapted to operate as a mobile high-voltage application system. The primary design goal was to ensure autonomous power operation via a portable AC generator, capable of handling dual-voltage input (110 or 220 V AC) using an integrated step-down transformer. Mechanically, a wheeled cart structure was developed to facilitate transportation across constrained or irregular terrains, such as abandoned lots, inoperative swimming pools, and disused water reservoirs (Fig. 10.7).

Electrically, the system received several important upgrades. A manual control panel was integrated, providing real-time feedback through voltage and current instrumentation, and including protective circuit breakers to safeguard against overloads or internal failure. The original trigger mechanism was replaced with a dual-button serial activation logic system, ensuring redundancy and minimizing the risk of accidental actuation. The resistive output load was reduced to 500 kΩ to accommodate noncontinuous high-voltage discharges, enabling short, safe energy bursts even under open-circuit conditions. The high-voltage transformer and power resistors were physically segregated and mounted in insulated sections within the mobile structure, ensuring compliance with safety requirements and avoiding accidental contact. Overall, the system design focused on safe high-voltage operation in uncontrolled outdoor environments.

Test protocols and procedures

Initial system characterization was performed with resistive dummy loads. These tests ensured thermal, electrical, and operational stability across the expected load range. Following the validation of electrical behavior under resistive conditions, the system was tested in water-filled containers simulating real breeding environments of *Aedes aegypti* (Fig. 10.8).

Test procedures included controlled variation of water volume and salinity to reflect real-world differences in natural stagnant water sites. Sodium chloride (NaCl) was added to increase conductivity where necessary. Two primary voltage levels—127 and 220 V AC—were used to simulate minimal and full-power scenarios. The primary evaluation parameter was mortality rate of larvae and pupae posttreatment. Observations extended for up to 72 h post-exposure to account for delayed effects caused by sublethal electrophysical damage. Primary voltages have, approximately, a square root effect on delivered power—110 V would end up delivering about one quarter of what could be delivered by a 220 V AC input.

Biological efficacy tests

In the first test (Fig. 10.9), a 15-liter container filled with stagnant water hosted 10 live *Aedes aegypti* larvae. The Fulgur system operated at 209 V AC, drawing 5.9 A, with an estimated output voltage of 1,390 V and a total power consumption of approximately 1,700 W. The discharge lasted 90 s. Eight larvae were killed instantly. The remaining two exhibited severe neuromuscular impairment and died within 72

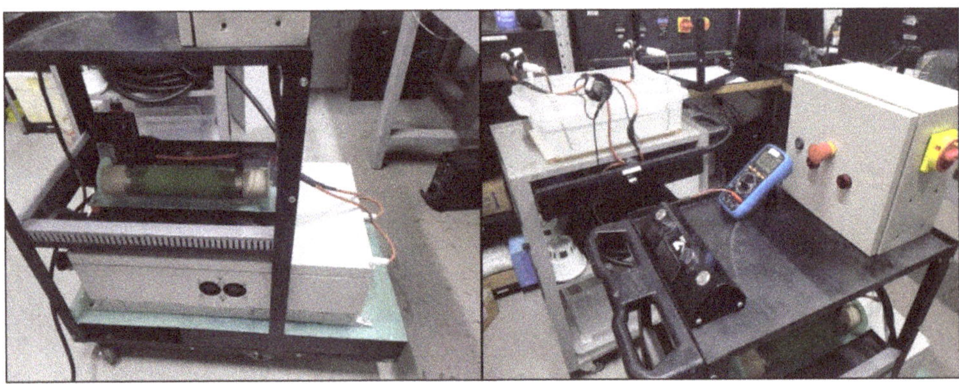

Fig. 10.7. Fulgur setup.

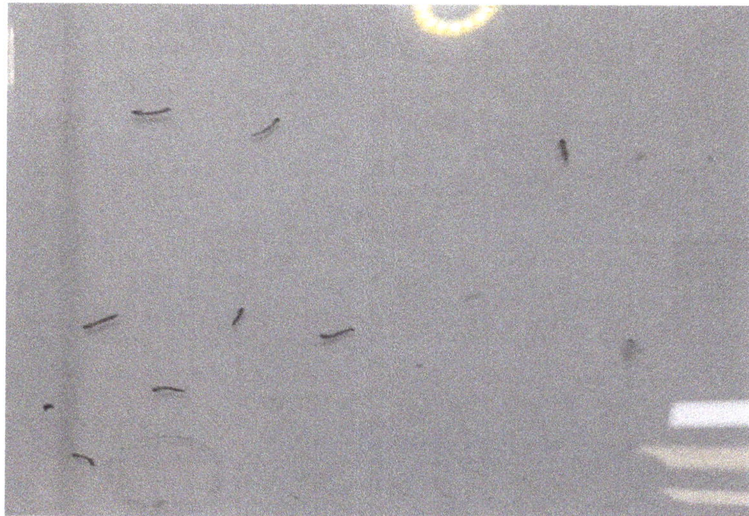

Fig. 10.8. *Aedes* larvae.

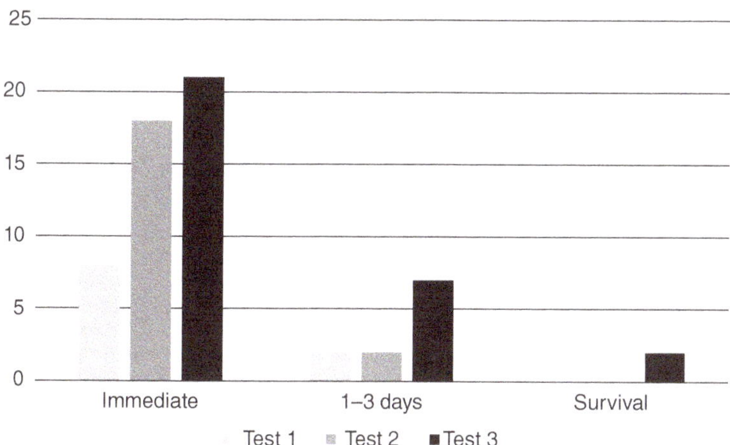

Fig. 10.9. Effects on larvae: test results.

h. Full mortality was confirmed by the end of the observation window.

In the second test, a larger 20-liter container held approximately 40 live organisms, including both larvae and pupae. The input at the primary side was 208 V AC at 6.3 A, resulting in a 1,470-volt output and a power level of roughly 1,770 W. Again, the discharge lasted 90 s. Thirty-five organisms, mostly larvae, died immediately. Two larvae and three pupae died within the next three days. The overall result was full mortality of all exposed individuals within 72 h.

The third test used a natural water sample of 10 l collected from an outdoor site on the premises of Zasso. This sample contained approximately 30 organisms, mostly pupae, in unknown stages of development. The system parameters matched the previous test. Mortality was observed at approximately 70% immediately following the discharge. Remaining organisms showed signs of trauma and immobility. By the following week, two of the survivors had reached adulthood. No taxonomic confirmation of the survivors' species was possible due to deterioration and lack of entomological facilities. Despite this, the results reinforced the need for operational adjustments to maintain full efficacy in naturally variable conditions.

Technical discussion

The system's performance was highly sensitive to the electrical conductivity of the treated water. Increased salinity reduced impedance, improving power transfer and maximizing current delivery to the larval targets. The 500-kilo-ohm output resistor, combined with real-time voltage regulation, ensured energy delivery remained within safe operational limits while maintaining efficacy.

The biological mechanism of action is assumed to include a combination of electroporation, electrolysis, and thermal effects. Electroporation likely disrupts cellular membrane integrity, causing irreversible damage. Electrolysis may contribute to pH shifts and localized chemical imbalances. Thermal effects appeared to be minimal under short application durations, though likely contributed to cumulative physiological stress. Delayed mortality in some specimens suggested that sublethal exposure can still result in complete developmental arrest or secondary infection leading to death. The ability to deliver lethal or near-lethal damage without physical contact is a unique advantage of the electrical approach over conventional larvicides.

Conclusion

The Fulgur module, in its mobile high-voltage configuration, demonstrated strong potential as a chemical-free solution for eliminating *Aedes aegypti* larvae and pupae in urban water accumulations. Mortality rates exceeded 90% in all controlled scenarios, with complete eradication achieved in low- and medium-volume test conditions. Even under natural field conditions, efficacy remained above 70%, validating the system's operational viability.

Future enhancements should include waveform modulation to increase the biological impact of each pulse. Ripple enhancement or the use of quasi-square waveforms may significantly improve electrocution efficiency. Digital integration with a microcontroller system will allow for precision timing, application consistency, and automated safety checks. Broader field trials across different ecological regions and with varying species of mosquitoes are recommended to establish universal applicability.

The proposed next-generation platform should be rugged, portable, and suitable for deployment by municipal health departments, environmental agencies, and emergency response teams.

Nematodes

Controlling nematodes in agriculture is crucial for several reasons, with one of the most significant being the prevention of lost production. Nematodes are microscopic roundworms that can infect plant roots, leading to various detrimental effects such as stunted growth, reduced yield, and even plant death.

Nematodes feed on plant roots, causing damage that impairs the plant's ability to absorb water and nutrients from the soil. This results in reduced plant growth and ultimately leads to lower yields at harvest time.

Nematode-infested plants often exhibit symptoms of stunted growth, including smaller leaves, fewer flowers, and overall weaker plant structure. This not only decreases the quantity of harvested produce but also diminishes its quality.

Severe nematode infestations can lead to significant crop losses. In some cases, entire fields may become unproductive or economically unviable due to nematode damage. This can have devastating consequences for farmers, leading to financial hardship and potentially threatening food security.

Farmers may incur additional expenses to manage nematode populations, such as purchasing and applying chemical nematicides or implementing crop rotation strategies. These increased production costs can further reduce profitability and competitiveness in the agricultural market.

Chemical nematicides used to control nematodes can have adverse effects on the environment, including soil and water contamination, as well as harm to nontarget organisms. Sustainable nematode control methods, such as biological control or resistant crop varieties, can help mitigate these environmental risks.

In summary, effective nematode control in agriculture is essential for maintaining high crop yields, ensuring food security, and promoting sustainable farming practices. By minimizing the impact of nematode damage, farmers can maximize their productivity and profitability while reducing environmental risks associated with conventional control methods.

As an example of the nematode impact in agriculture, please see Fig. 10.10 and Fig. 10.11:

Electricity can be used not only to control invasive plants, but also to control nematodes. The following is based in the work done by UNESP FCAV Campus Jaboticabal, Depto. Fitossanidade–LABNEMA, January–April 2021, and was presented at the 37th Brazilian Nematology Congress in 2022. The objective was to study the efficiency of electrical discharge in controlling nematodes in soybean and sugarcane:

- soybean (*Meloidogyne incognita*)
- sugarcane (*Pratylenchus zeae* and *M. javanica*)

Randomized experimental design:

- soybean: 5 treatments and 4 replications (20 plots)
- sugarcane: 2 treatments and 4 replications (8 plots)

Inoculation:

- Soybean: one plant per pot, autoclaved substrate, inoculated with 1,000 eggs + second-stage juveniles.
- Sugarcane: one seedling per pot, containing soil naturally infested with *Pratylenchus zeae* and *M. javanica*.

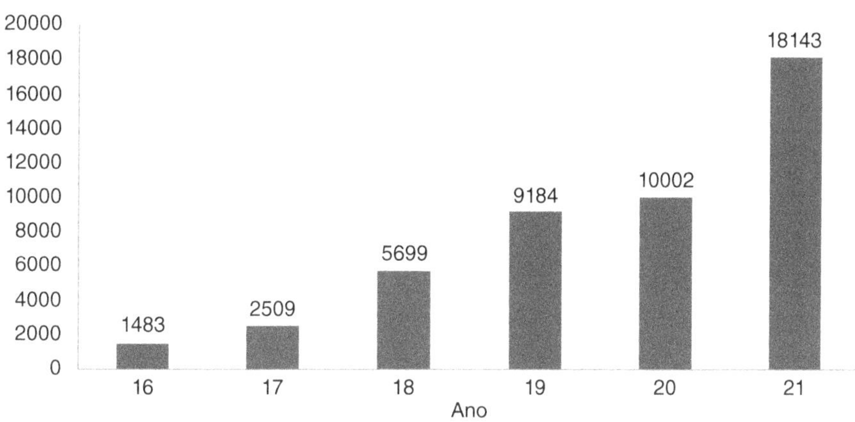

Fig. 10.10. Evolution in the Brazilian market of nematicides (in 1,000 ha).

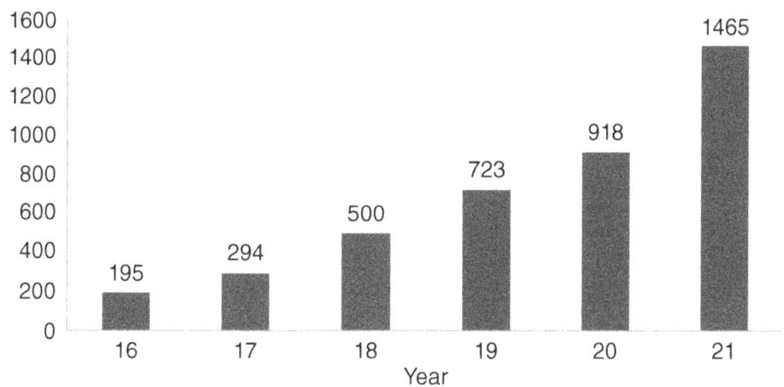

Fig. 10.11. Evolution of the Brazilian nematicide market (in R$1,000,000).

Electrical discharge was applied by Zasso Brasil equipment. The positive electrode was attached to the plant stem, 15 cm above the ground, while the negative electrode was inserted into the soil (Fig. 10.12 and Fig. 10.13):

Fig. 10.12. Plants in vase for research setup.

Using soybean, all tested electrical discharges significantly reduced the final population of active juveniles of *M. incognita* in soybean roots compared to the control. It should be noted that the plots that received the treatment with 4 s of exposure, showed an estimated population reduction of *M. incognita* of 90%, which is superior to any known allowed method (Fig. 10.14).

In sugarcane, the applied electrical discharge reduced the population of active juveniles and adults of *Pratylenchus zeae* and juveniles of *M. javanica* present in the sugarcane roots after electrical discharge application compared to the control, showing an estimated population reduction of *Pratylenchus zeae* by 88% and *M. javanica* by 83% (Table 10.1).

In the follow-up experiment the objective was to evaluate the efficacy of electrocution doses in soybeans for nematode control.

- **Control.** TAP1, TAP2, TAP3, and TAP4—levels of electrical energy applied.
- **Experimental Design.** The experiment was conducted at Fazenda Cachoerinha, Guaíra, Sao Paulo, Brazil, with soybean plants at stage R5 grown in an area with high nematode infestation (*M. javanica, M. incognita, Pratylenchus brachyurus,* and *Heterodera glycines*).

Fig. 10.13. Apparatus used in the modified Baermann method (1917) for extraction of active (live) nematodes.

Soybean planting rows were electrocuted with four different doses (TAP1 to TAP4) of electrocution, in addition to the control. One day after electrocution, plants showing symptoms of nematode infection were collected, and nematode multiplication readings were taken 68 days after collection and inoculation in pots.

In conclusion, electrocution can reduce up to 86.1% of the total nematode population in the roots and up to 92.1% of the nematode population in the soil, compared to the control, without causing any harm to the development of the soybean plant (Fig. 10.15, Fig. 10.16, and Fig. 10.17).

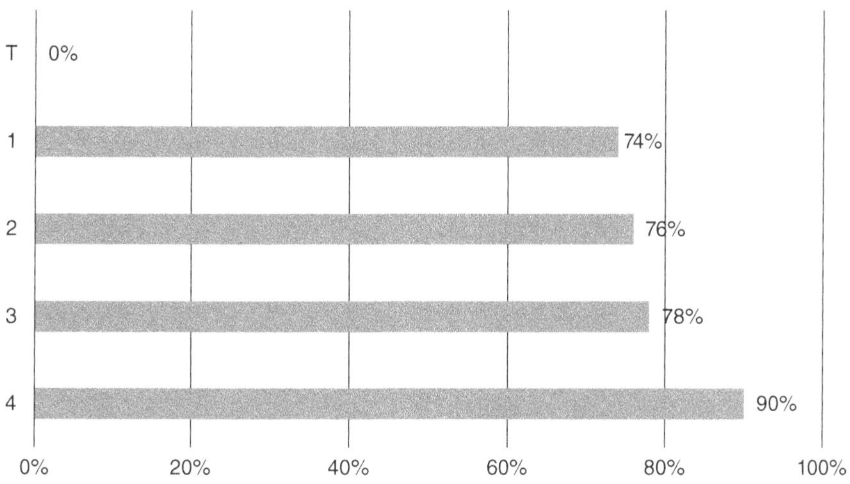

Fig. 10.14. Efficiency in control (E%) in different times of application of electricity.

Table 10.1. Nematode effects: before and after application.

Test	*Pratylenchus zeae*	Control (%)	*Meloidogyne javanica*	Control (%)
Control	27.294	0	4.161	0
1s	3.174	88	688	83

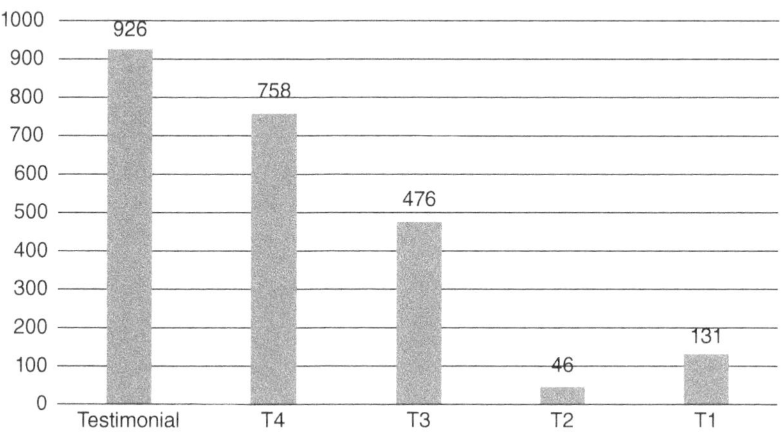

Fig. 10.15. Soil after treatment: nematodes (#/100 cm^3).

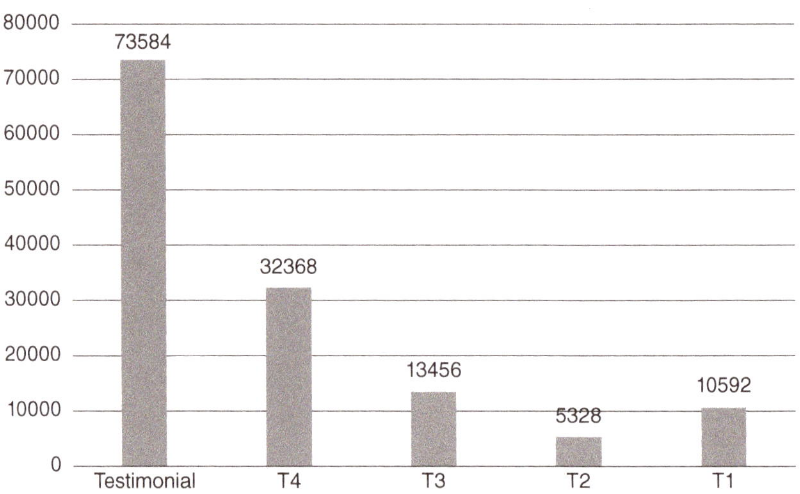

Fig. 10.16. Root after treatment: nematodes (#/10 g).

Fig. 10.17. Pictures of the field and plants treated at different power levels.

Cotton Stems

In the *Destruction of Cotton Stubble with Electrical Weeding. Primavera do Leste – MT, January 2024*, by Autieres Teixeira Faria and Luísa Limberger Mrozinski (2024), conducted in Primavera do Leste, Mato Grosso, Brazil (January 2024), the results demonstrate the efficiency of electrical weeding for cotton stubble destruction under the given soil and climatic conditions. Key findings indicate:

- **Effective stubble destruction.** The application of electrical discharge successfully eliminated cotton stubble from the first evaluation stage, making a second application unnecessary.
- **Superior performance compared to chemical control.** Electrical weeding outperformed chemical herbicide treatments, particularly in early-stage assessments, with significantly fewer plant regrowth instances.
- **Enhanced agronomic and environmental benefits.** The near-total elimination of treated plants reduces the risk of pest and disease proliferation in the following crop cycle, supporting more sustainable agricultural management.

These findings reinforce the viability of electrical weeding as a practical and eco-friendly alternative to traditional chemical control methods for managing postharvest cotton residues (Table 10.2).

Results

The untreated control group consistently has the highest plant regrowth across all time points (2.90 plants m^{-2} at 21 days after application [DAA]), indicating that natural persistence remains strong without intervention.

Among the electrical treatments, high and medium voltage resulted in the lowest regrowth across all time points (0.35 plants m^{-2} at 7 DAA, 0.05 at 14 DAA, and 0.10 at 21 DAA), suggesting it is the most effective electrical treatment.

The effectiveness of 2,4-D varies depending on when it was applied. Earlier applications (7 or 14 DAA) resulted in higher regrowth (2.80–2.60 plants m^{-2} at 7 DAA) compared to later application (21 days after planting [DAP]: 2.35 plants m^{-2}), indicating that timing plays a crucial role in herbicide effectiveness. After 21 days, regrowth is significantly reduced in all cases but not as effectively as some electrical treatments.

Electrical weeding, particularly at medium voltage, demonstrated superior long-term effectiveness compared to chemical herbicides and the control group. Overall, electrical weeding appears to be a promising alternative to chemical herbicides, especially at optimized voltage levels.

Table 10.2. Regrowth per treatment type and time.

	Regrown plants m^{-2}		
Treatments	7 days after application (DAA)	14 DAA	21 DAA
2,4-D 21 DAP	2.35	1.45	1.00
Weak voltage	0.95	0.25	0.40
Medium voltage	0.35	0.05	0.10
Strong voltage	1.10	1.00	0.90
Strong voltage	1.25	0.05	0.15
2,4-D 14 DAA	2.60	1.70	0.60
2,4-D 7 DAA	2.80	1.85	1.15
Control	2.90	2.70	2.90

Similar study

These results were replicated by MT Foundation, in research work done by Lucas Heringer Barcellos Júnior at Rondonópolis, Mato Grosso, Brazil. The results are shown in Table 10.3:

This study evaluates different treatments for controlling plant regrowth using chemical (2,4-D+Oil) and electrical (Electroherb™) methods. The key observations are:

- **Chemical control (2,4-D+Oil).** Applied in two phases, it provided consistent treatment with measured doses, requiring multiple applications to ensure effectiveness.
- **Electrical control (Electroherb™).** Different voltage levels (weak, medium, and

Table 10.3. Regrowth per treatment type and dose.

Herbicide	Application timing	Dose (l/ha)	Dose (ml) per 1.8 l
2,4-D+Oil	1st application (apply to the stump up to 30 min after mowing)	2+0.5	30+7.5
2,4-D+Oil	2nd application (apply to regrowth up to 4 leaves)	2+0.5	30+7.5
Electroherb™	1st application (apply to the stump up to 30 min after mowing)	Weak voltage	-
Electroherb™	2nd application (apply to regrowth up to 4 leaves)	Weak voltage	-
Electroherb™	1st application (apply to the stump up to 30 min after mowing)	Medium voltage	-
Electroherb™	2nd application (apply to regrowth up to 4 leaves)	Medium voltage	-
Electroherb™	1st application (apply to the stump up to 30 min after mowing)	Strong voltage	-
Electroherb™	2nd application (apply to regrowth up to 4 leaves)	Strong voltage	-
Electroherb™+Avicta	Seed treatment	100 ml/100 kg	-
2,4-D+Oil	1st application (apply to the stump up to 30 min after mowing)	2+0.5	30+7.5
2,4-D+Oil	2nd application (apply to regrowth up to 4 leaves)	2+0.5	30+7.5
2,4-D+Oil+Avicta	Seed treatment	100 ml/100 kg	-

strong) were tested. Higher voltage applications showed superior performance in eliminating regrowth, reducing the need for additional interventions.

- **Seed treatment (Avicta Complete).** Used in some treatments as a complementary measure, ensuring broader pest control in treated seeds.

Overall, electrical weeding demonstrated a promising alternative to chemical herbicides, with strong voltage levels showing efficiency comparable or superior to traditional chemical treatments.

The electrical application was carried out with a prototype developed by Zasso, which involved passing metal plates directly in contact with the cotton plant stubble. The initial plan was for the equipment to pass twice, simulating a sequential electric discharge. However, the second pass was not necessary, as the stubble was already dead.

Chemical applications were carried out using the herbicide 2,4-D in two applications, as shown in Table 10.4.

Although the protocol included two periods of electrical discharge application, the second one was not necessary. The percentage of control in the evaluated periods was high in all treatments. However, treatments with electricity stood out, being superior to chemical control regardless of the applied voltage levels. Only from 21 DAA were the treatments statistically equal. The weight of 1,000 seeds and soybean productivity did not differ statistically among the treatments, averaging 199.8 grams and 58 sacks/ha, respectively.

Up to 28 DAA, treatments with chemical destruction of stubble showed an index of 2–15% of plants with regrowth, which did not occur with treatments using electrical discharge.

Controlling cotton stems in agriculture is crucial to mitigate the impact of pests such as the "Bicudo" or boll weevil (*Anthonomus grandis*) for several reasons:

- **Economic impact.** Cotton is a significant cash crop grown for its fiber, which is used in various textile products. The presence of pests like the Bicudo can lead to significant economic losses for cotton farmers. These losses occur due to reduced yields, lower fiber quality, and increased production costs associated with pest management.
- **Quality degradation.** Pests like the Bicudo not only reduce cotton yields but also affect the quality of the fiber. Damage to cotton bolls can result in stained lint, decreased fiber length and strength, and increased trash content. Lower-quality cotton fiber fetches lower prices in the market, further exacerbating economic losses for farmers.
- **Increased production costs.** Controlling pests like the Bicudo often requires the use of insecticides and other management practices, which add to the production costs for cotton farmers. These costs include purchasing insecticides, application equipment, and labor for pest monitoring and control activities. Moreover, repeated

Table 10.4. Method of treatment.

Method	l/ha	Details
2,4-D+oil	2+0.5	1st application 30 min after mowing, 2nd after regrowth up to 4 leaves
Electroherb™	Low power	1st application 30 min after mowing, 2nd after regrowth up to 4 leaves
Electroherb™	Mid power	1st application 30 min after mowing, 2nd after regrowth up to 4 leaves
Electroherb™	High power	1st application 30 min after mowing, 2nd after regrowth up to 4 leaves
EH + Avicta	HP + Avicta (1 ml/kg)	1st application 30 min after mowing, 2nd after regrowth up to 4 leaves
2,4-D+oil	2+0.5	1st application 30 min after mowing
2,4-D+oil	2+0.5	2nd after regrowth up to 4 leaves
Avicta	Avicta (1 ml/kg of seeds)	Just Avicta

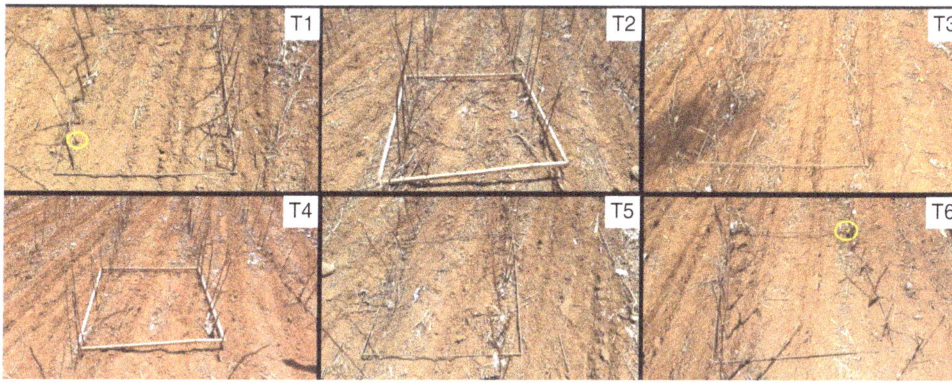

Fig. 10.18. Photos of the treatments at 7 days after application. Regrowth is highlighted by a ring in treatments T1 and T6.

Fig. 10.19. Photos of the treatments at 10 days after soybean planting.

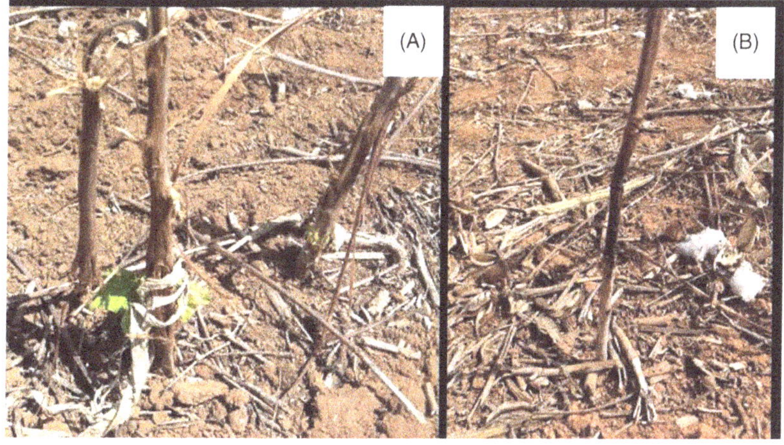

Fig. 10.20. Highlight of stubble with regrowth (A) and visual aspect of the stubble after electricity (B).

Fig. 10.21. Prototype at the time of application (A) and detail of the metal plates in contact with the cotton plant stubble (B).

insecticide applications can lead to the development of pesticide resistance in pest populations, necessitating the use of more expensive or environmentally harmful chemicals.

- **Environmental concerns.** Pesticide use in cotton production can have adverse effects on the environment, including soil and water contamination, harm to nontarget organisms, and disruption of ecosystem balance. By controlling cotton pests through sustainable and integrated approaches, farmers can minimize the environmental impact of pest management practices.

In summary, controlling cotton stems in agriculture is essential for maintaining cotton yield and quality, minimizing production costs, and reducing the environmental impact of pest management activities. Effective pest control measures help ensure the sustainability and profitability of cotton farming operations while safeguarding the livelihoods of cotton farmers. With that in mind, a successful attempt to control cotton stems was made with the use of electricity (Fig. 10.18, Fig. 10.19, Fig. 10.20, and Fig. 10.21).

Results

The conclusions were clear. Under the soil and climate conditions of this study, the results obtained allow us to consider that electrical discharge was effective in destroying cotton stubble from the first evaluation period, control with electricity was superior to chemical control in the initial evaluations, and electrical discharge on cotton stubble did not affect the weight of 1,000 seeds and soybean productivity.

References

Coutinho Filho, S.A. and Rinzler, G.P.M. (2022) *Aplicador, Sistema e Método para Aplicar Corrente Elétrica em uma Planta, e, Pivô Agrícola. Patent No. PCT/BR2022/050406*. World Intellectual Property Organization (WIPO), Geneva, Switzerland.

Faria, A.T. and Mrozinski, L.L. (2024) *Destruction of Cotton Stubble with Electrical Weeding*. Lucrativa Consultoria, Primavera do Leste, Brazil.

Zasso, A.G. and SENAI (2024) *Sistema e Método de Sensoriamento para Identificação de Plantas Invasoras. Patent No. WO 2024/234068*. World Intellectual Property Organization (WIPO), Geneva, Switzerland.

Index